LONDON MATHEMATICAL SOCIETY LECTURE NOTE SERIE.

Managing Editor: Professor J.W.S. Cassels,
Department of Pure Mathematics and Mathematical Statistics,
16 Mill Lane, Cambridge CB2 1SB.

1. General cohomology theory and K-theory, P.HILTON
4. Algebraic topology, J.F.ADAMS
5. Commutative algebra, J.T.KNIGHT
8. Integration and harmonic analysis on compact groups, R.E.EDWARDS
9. Elliptic functions and elliptic curves, P.DU VAL
10. Numerical ranges II, F.F.BONSALL & J.DUNCAN
11. New developments in topology, G.SEGAL (ed.)
12. Symposium on complex analysis, Canterbury, 1973, J.CLUNIE & W.K.HAYMAN (eds.)
13. Combinatorics: Proceedings of the British Combinatorial Conference 1973, T.P.McDONOUGH & V.C.MAVRON (eds.)
15. An introduction to topological groups, P.J.HIGGINS
16. Topics in finite groups, T.M.GAGEN
17. Differential germs and catastrophes, Th.BROCKER & L.LANDER
18. A geometric approach to homology theory, S.BUONCRISTIANO, C.P. ROURKE & B.J.SANDERSON
20. Sheaf theory, B.R.TENNISON
21. Automatic continuity of linear operators, A.M.SINCLAIR
23. Parallelisms of complete designs, P.J.CAMERON
24. The topology of Stiefel manifolds, I.M.JAMES
25. Lie groups and compact groups, J.F.PRICE
26. Transformation groups: Proceedings of the conference in the University of Newcastle-upon-Tyne, August 1976, C.KOSNIOWSKI
27. Skew field constructions, P.M.COHN
28. Brownian motion, Hardy spaces and bounded mean oscillations, K.E.PETERSEN
29. Pontryagin duality and the structure of locally compact Abelian groups, S.A.MORRIS
30. Interaction models, N.L.BIGGS
31. Continuous crossed products and type III von Neumann algebras, A.VAN DAELE
32. Uniform algebras and Jensen measures, T.W.GAMELIN
33. Permutation groups and combinatorial structures, N.L.BIGGS & A.T.WHITE
34. Representation theory of Lie groups, M.F. ATIYAH et al.
35. Trace ideals and their applications, B.SIMON
36. Homological group theory, C.T.C.WALL (ed.)
37. Partially ordered rings and semi-algebraic geometry, G.W.BRUMFIEL
38. Surveys in combinatorics, B.BOLLOBAS (ed.)
39. Affine sets and affine groups, D.G.NORTHCOTT
40. Introduction to Hp spaces, P.J.KOOSIS
41. Theory and applications of Hopf bifurcation, B.D.HASSARD, N.D.KAZARINOFF & Y-H.WAN
42. Topics in the theory of group presentations, D.L.JOHNSON
43. Graphs, codes and designs, P.J.CAMERON & J.H.VAN LINT
44. Z/2-homotopy theory, M.C.CRABB
45. Recursion theory: its generalisations and applications, F.R.DRAKE & S.S.WAINER (eds.)
46. p-adic analysis: a short course on recent work, N.KOBLITZ
47. Coding the Universe, A.BELLER, R.JENSEN & P.WELCH
48. Low-dimensional topology, R.BROWN & T.L.THICKSTUN (eds.)

49. Finite geometries and designs, P.CAMERON, J.W.P.HIRSCHFELD & D.R.HUGHES (eds.)
50. Commutator calculus and groups of homotopy classes, H.J.BAUES
51. Synthetic differential geometry, A.KOCK
52. Combinatorics, H.N.V.TEMPERLEY (ed.)
53. Singularity theory, V.I.ARNOLD
54. Markov processes and related problems of analysis, E.B.DYNKIN
55. Ordered permutation groups, A.M.W.GLASS
56. Journées arithmétiques 1980, J.V.ARMITAGE (ed.)
57. Techniques of geometric topology, R.A.FENN
58. Singularities of smooth functions and maps, J.MARTINET
59. Applicable differential geometry, M.CRAMPIN & F.A.E.PIRANI
60. Integrable systems, S.P.NOVIKOV et al.
61. The core model, A.DODD
62. Economics for mathematicians, J.W.S.CASSELS
63. Continuous semigroups in Banach algebras, A.M.SINCLAIR
64. Basic concepts of enriched category theory, G.M.KELLY
65. Several complex variables and complex manifolds I, M.J.FIELD
66. Several complex variables and complex manifolds II, M.J.FIELD
67. Classification problems in ergodic theory, W.PARRY & S.TUNCEL
68. Complex algebraic surfaces, A.BEAUVILLE
69. Representation theory, I.M.GELFAND et al.
70. Stochastic differential equations on manifolds, K.D.ELWORTHY
71. Groups - St Andrews 1981, C.M.CAMPBELL & E.F.ROBERTSON (eds.)
72. Commutative algebra: Durham 1981, R.Y.SHARP (ed.)
73. Riemann surfaces: a view towards several complex variables, A.T.HUCKLEBERRY
74. Symmetric designs: an algebraic approach, E.S.LANDER
75. New geometric splittings of classical knots (algebraic knots), L.SIEBENMANN & F.BONAHON
76. Linear differential operators, H.O.CORDES
77. Isolated singular points on complete intersections, E.J.N.LOOIJENGA
78. A primer on Riemann surfaces, A.F.BEARDON
79. Probability, statistics and analysis, J.F.C.KINGMAN & G.E.H.REUTER (eds.)
80. Introduction to the representation theory of compact and locally compact groups, A.ROBERT
81. Skew fields, P.K.DRAXL
82. Surveys in combinatorics: Invited papers for the ninth British Combinatorial Conference 1983, E.K.LLOYD (ed.)
83. Homogeneous structures on Riemannian manifolds, F.TRICERRI & L.VANHECKE
84. Finite group algebras and their modules, P.LANDROCK
85. Solitons, P.G.DRAZIN
86. Topological topics, I.M.JAMES (ed.)
87. Surveys in set theory, A.R.D.MATHIAS (ed.)
88. FPF ring theory, C.FAITH & S.PAGE
89. An F-space sampler, N.J.KALTON, N.T.PECK & J.W.ROBERTS
90. Polytopes and symmetry, S.A.ROBERTSON
91. Classgroups of group rings, M.J.TAYLOR
92. Simple artinian rings, A.H.SCHOFIELD
93. General and algebraic topology, I.M.JAMES & E.H.KRONHEIMER
94. Representations of general linear groups, G.D.JAMES

London Mathematical Society Lecture Note Series: 91

Classgroups of Group Rings

MARTIN TAYLOR

Fellow of Trinity College, Cambridge

The right of the University of Cambridge to print and sell all manner of books was granted by Henry VIII in 1534. The University has printed and published continuously since 1584.

CAMBRIDGE UNIVERSITY PRESS

Cambridge

London New York New Rochelle

Melbourne Sydney

CAMBRIDGE UNIVERSITY PRESS
Cambridge, New York, Melbourne, Madrid, Cape Town, Singapore, São Paulo

Cambridge University Press
The Edinburgh Building, Cambridge CB2 8RU, UK

Published in the United States of America by Cambridge University Press, New York

www.cambridge.org
Information on this title: www.cambridge.org/9780521278706

First published 1984
Re-issued in this digitally printed version 2008

A catalogue record for this publication is available from the British Library

Library of Congress Catalogue Card Number: 83–26167

ISBN 978-0-521-27870-6 paperback

To Sharon

CONTENTS

PAGE

PREFACE

It gives me great pleasure to thank Ali Fröhlich for all the help and encouragement he has given me, and in particular for the many suggestions and helpful remarks he has made concerning the writing of this book.

I should like to express my thanks to Steve Ullom for numerous conversations and valuable insights, and I also wish to thank Joan Bunn for the typing.

This book first appeared in the form of an essay submitted for the Adams prize at Cambridge University.

Trinity College
Cambridge
July 1983.

INTRODUCTION

For the most part this book is concerned with modules, which are locally free over an integral group ring, and the consequent problem of determining whether or not the module is globally free. Such questions arise naturally in both algebraic number theory and in algebraic topology. In the former, the standard such question is that of the existence of a normal integral basis. That is to say, given a Galois extension of number fields N/K, we wish to know whether $\mathcal{O}_N$, the ring of integers of N, has a basis over the ring of integers of a subfield $F \subset K$ which has the form $\{a_i^\gamma\}$, for γ running through Gal(N/K). An alternative way of considering this is to ask whether or not $\mathcal{O}_N$ is free over the group algebra $\mathcal{O}_F$ Gal(N/K). In the second area of application, that of algebraic topology, C.T.C. Wall has introduced a locally free module whose deviation from being globally free represents an obstruction to finding a finite complex in the homotopy type of a given space.

For a group ring R, $C\ell(R)$ is defined in a manner which closely resembles the way in which the ideal classgroup of a Dedekind domain is defined. In place of taking ideals modulo principal ideals, in essence we take locally free modules modulo free modules. This construction will be made precise in chapter 1.

The principal aim of this book is to instruct the reader in sufficient techniques to enable him, when given a locally free R module M, to calculate the class of M in $C\ell(R)$ and thereby, in many cases, determine whether or not M is globally free.

In the 1960's Swan and Jacobinski gave abstract descriptions of locally free (or projective) classgroups. However, explicit calculations of these classgroups were possible in only a few cases. Subsequently Reiner and Ullom introduced the elegant technique of Mayer-Vietoris sequences to the subject. This proved to be quite a powerful tool and it substantially increased our knowledge of classgroups of group rings. There is an excellent account of the level of knowledge obtained by such methods in S. Ullom's survey article [U3].

There then came a very important turning point in the theory of such classgroups when A. Fröhlich, motivated by the normal integral basis problem, introduced a completely new description of such classgroups. He was able to show that such classgroups are naturally isomorphic to the quotient of two groups of homomorphisms from the virtual characters of the group in question, taking idelic values. With this new viewpoint even the most basic properties were immediately better understood. As an example we consider certain elementary functorial properties. Previous descriptions of such classgroups had nearly always been in terms of families of ideals, one for each absolutely irreducible character of the underlying group. If we now change group, by means of induction resp. restriction of module, this then corresponds to restriction resp. induction on the characters of the underlying group. Of course, in general, induction and restriction do not preserve irreducibility of characters. They do, however, induce natural maps on homomorphisms from virtual characters. To underline the advantage of this change in viewpoint, we mention one further development. Presently we shall see that the Adams operations of character theory play a fundamental role in the theory of classgroups. In general, however, an Adams operation takes an irreducible character to a virtual character (and not an actual character, let alone an irreducible character).

Whilst Fröhlich's description of classgroups represents a fundamental change in view point, it does not, however, solve the basic problems. The point being that while the numerator in his description of the classgroup as the quotient of two groups was very clearly understood, the denominator was exceedingly difficult to handle. This necessitated a further development called the group logarithm which was first introduced by the author in [T1]. Essentially the group logarithm is the usual p-adic logarithm twisted by means of an Adams operation. Very often, adroit use of this logarithm, together with Fröhlich's description, allows us to decide whether or not the class of a module is trivial or not. The power of this technique is well illustrated by the normal integral basis problem. Here, the above work, when allied to Fröhlich's Gauss sum resolvent relation, enables one to describe the class of a ring of integers over a group ring whose coefficients are $\mathbb{Z}$, whenever the ring of integers is locally free (cf. [T3], [F7]). There are numerous other results which follow by applying this strategy: I mention in particular Ullom's conjecture on Swan classes of p-groups cf. Chapter 3, the fixed point theorem for group determinants cf. Chapter 8, Wall's conjecture for determinants concentrated at the identity cf. Chapter 7, and the existence of Adams operations on classgroups cf. Chapter 9.

Even though my own motivation for studying classgroups of group rings is overtly arithmetic, I have tried to present this work as a piece of pure algebra, in order that it be accessible to as wide an audience as possible. There is one point where this restraint gave way; namely, when describing the beautiful theory of Swan modules, I give a proof of the self-duality of a ring of integers as a Galois module (in the so-called tame case).

In order to illustrate these various new techniques, I have tried to include as large a number of worked examples as possible. In particular, it is worth mentioning that I have included the classes of Swan modules of p-groups, the classgroups of generalised dihedral pq groups (due to Galovich-Reiner-Ullom) and the so-called exceptional 2-groups, i.e. dihedral, semi-dihedral and quaternion 2-groups (due to Fröhlich-Keating-Wilson-Endo).

For the reader who wishes to go further, I should mention A. Fröhlich's two books 'Galois module structure of algebraic integers' and 'Classgroups, in particular Hermitian classgroups'. In the former there is a full account of the Normal Integral Basis problem, and in the latter a theory, parallel to that described here, is constructed for modules with a Hermitian form. Here, for the first time, I give a full account of the basic theory of locally free classgroups of group rings, together with the group logarithm and most of the fundamental results in the area.

1. FRÖHLICH'S DESCRIPTION OF CLASSGROUPS

§1. AN EXACT SEQUENCE FROM K-THEORY

Throughout all rings have a 1, and, for a given ring R, the multiplicative group of its invertible elements is denoted by R^*.

In general modules are taken to have their action on the right, and, unless stated otherwise, all modules are assumed to be finitely generated.

In this section we shall put the locally free classgroup in the context of K-theory. In particular, we describe its appearance in an exact sequence of K-theory. Our presentation is based on that given in chapter 1 of [F2].

Let K be a number field or a finite extension of the p-adic field Q_p. The ring of integers of K will be denoted by $\mathcal{O}_K$. Let A be a finite dimensional semi-simple K algebra and let $\mathcal{A}$ be an $\mathcal{O}_K$-order in A. The Grothendieck group of locally free $\mathcal{A}$ modules denoted $K_o(\mathcal{A})$, is the free abelian group on the isomorphism classes of locally free $\mathcal{A}$-modules, modulo the group generated by the relations [N] - [M] - [P] for each exact sequence of locally free $\mathcal{A}$-modules

$$0 \to M \to N \to P \to 0.$$

It is worth remarking that this Grothendieck group is more restrictive than the usual one where all projective $\mathcal{A}$ modules are considered. However, in the crucial case of a global group ring, Swan has shown

Theorem 1.1 [Sw2]

Let K be a number field, let Γ be a finite group and let M be an $\mathcal{O}_K\Gamma$ module. Then M is projective over $\mathcal{O}_K\Gamma$ if, and only if, it is a locally free $\mathcal{O}_K\Gamma$-module.

An $\mathcal{A}$-module M is said to be locally freely presented if there exists an exact sequence of Λ-modules

$$0 \to P \to N \to M \to 0 \qquad (1.2)$$

with P and N locally free Λ-modules of the same rank. [Thus M is finite].

Remark If $\Lambda = \mathbb{Z}\Gamma$, then by Theorem 1.1, together with Rim's Theorem in [Ri], it follows that M is locally freely presented precisely when M is a finite $\mathbb{Z}\Gamma$ module which is cohomologically trivial.

Let $K_oT(\Lambda)$ denote the Grothendieck group of the category of such modules, taken with respect to exact sequences.

For a locally free Λ-module M (resp. locally freely presented Λ-module M) we shall write [M] for its class in $K_o(\Lambda)$ (resp. $K_oT(\Lambda)$).

There is a natural homomorphism

$$\delta\colon K_oT(\Lambda) \to K_o(\Lambda)$$

where for M, N, P as given in (1.2)

$$\delta([M]) = [P] - [N].$$

Recall that for a ring R,

$$K_1(R) = \frac{\lim\limits_{\to} GL_n(R)}{\lim\limits_{\to} E_n(R)}$$

where $E_n(R)$ denotes the group generated by the elementary matrices in $M_n(A)$.

From §I in [F2] we have a long exact sequence of K-theory

$$K_1(A) \xrightarrow{\kappa} K_1(A) \xrightarrow{\theta} K_oT(A) \xrightarrow{\delta} K_o(A) \xrightarrow{r} \mathbb{Z} \to 0 \qquad (1.3)$$

If one replaces locally free modules by projective modules in defining the above Grothendieck groups then the above becomes the well-known exact sequence of K-theory of Bass cf. [B] IX 6.3.

We now explain the homomorphisms used in (1.3): r is the homomorphism induced by associating to each locally free Λ-module its

A-rank, and κ is induced by the inclusion $A \hookrightarrow A$. The homomorphism θ is induced by the homomorphisms

$$\theta_n: GL_n(A) \to K_oT(A)$$

which are defined as follows. For $\alpha \in GL_n(A)$ we choose a locally free A module X which spans $\overset{n}{\underset{1}{\oplus}} A$ and some $c \in \mathcal{O}_K \setminus \{0\}$ such that $c X \subset X \cap x\alpha$; then

$$\theta_n(\alpha) = \left[\frac{X}{Xc}\right] - \left[\frac{X\alpha}{Xc}\right].$$

The locally free classgroup of A, $Cl(A)$ is defined to be Ker(r). It is worth remarking that if K is a finite extension of Q_p, then of course $Cl(A) = \{1\}$.

In general there is a homomorphism $K_o(A) \to Cl(A)$, where for a locally free A-module M, [M] maps to $[M] - rk_A(M)\ [A]$. We shall write (M) for this class of M in $Cl(A)$.

$Cl(A)$, and therefore by necessity $K_oT(A)$, are the main subject of this study, in the case when A is a group ring.

Let M be a maximal order of A containing A, then $\otimes_A M$ induces a homomorphism (which is shown to be a surjection in §3)

$$Cl(A) \to Cl(M).$$

We denote the kernel of this homomorphism by $D(A)$. It is well-known that in fact $D(A)$ is independent of the particular choice of maximal order M (cf. (3.10)). In §3, it will be shown that $Cl(M)$ is a product of certain ideal classgroups - a result which is originally due to Eichler. A great deal is known about such classgroups, and so, for the most part, we restrict our attention to $D(A)$. Even when $D(A)$ and $Cl(M)$ are both known, there remains the delicate question of describing the extension

$$1 \to D(A) \to Cl(A) \to Cl(M) \to 1$$

The only results in this direction that I am aware of are [U5] and [Mc].

§2. NOTATION

For a field F, let F^c be an algebraic closure of F. Once and for all we fix algebraic closures Q^c, Q_p^c for all rational primes p. Let $\mathbb{R}$ (resp. $\mathbb{C}$) be the field of real numbers (resp. field of complex numbers). We shall frequently speak of the Archimedean valuations of a number field as the infinite primes. With this convention $Q_\infty = \mathbb{R}$, $Q_\infty^c = \mathbb{C}$.

If K is a number field (resp. a finite extension of Q_p), then we let $\Omega_K = \mathrm{Gal}(Q^c/K)$ (resp. $\Omega_K = \mathrm{Gal}(Q_p^c/K)$).

(2.1) Let $K \subset Q^c$ be a number field, suppose $p < \infty$ and let $h: K \hookrightarrow Q_p^c$ be a field embedding. Let P^c (resp. U_p) be the maximal ideal (resp. group of units) in the ring of integers Q_p^c.

Let $\mathcal{p}$ denote the maximal ideal $h^{-1}(P^c)$. $\mathrm{Gal}(Q_p^c/Q_p)$ acts on such embeddings by composition, and the orbits (or by abuse of language a representative of an orbit) is called a place of K. In this way a bijection between places with $p < \infty$ and the maximal ideals of $\mathcal{O}_K$ is obtained. Alternatively, relaxing the condition $p < \infty$, we obtain a bijection between the places of K and the equivalence classes of valuations of K.

For a place h of K associated to the (possibly infinite) prime $\mathcal{p}$, let $K_{\mathcal{p}}$ be the closure of h(K) in Q_p^c (so that $K_{\mathcal{p}}$ is defined only up to Galois conjugacy). With the notation of the previous section, let $A_{\mathcal{p}}$ resp. $\mathcal{A}_{\mathcal{p}}$ denote the closure of the image of A (resp. $\mathcal{A}$) under the K-linear extension of h to $A \hookrightarrow A \otimes_K Q_p^c$.

For a rational prime p we write A_p (resp. $\mathcal{A}_p$) for $A \otimes_Q Q_p$ (resp. $\mathcal{A} \otimes_{\mathbb{Z}} \mathbb{Z}_p$). Then the isomorphism

$$K \otimes_Q Q_p \cong \prod_{\mathcal{p}|p} K_{\mathcal{p}}$$

induces isomorphisms $A_p \cong \prod A_{\mathcal{p}}$, $\mathcal{A}_p \cong \prod \mathcal{A}_{\mathcal{p}}$.

Let R_A denote the Grothendieck group of $A \otimes_K K^c$ modules. In the particular case when Γ is a finite group, $A = K\Gamma$, we write R_Γ (resp. $R_{\Gamma,p}$) for R_A when K is a number field (resp. a finite extension of Q_p). An embedding $Q^c \hookrightarrow Q_p^c$ determines an isomorphism $R_\Gamma \cong R_{\Gamma,p}$. Frequently such an embedding will be assumed to be given and $R_{\Gamma,p}$ and R_Γ will be identified.

Alternatively R_Γ (resp. $R_{\Gamma,p}$) will frequently be viewed as the group of virtual Q^c (resp. Q_p^c) characters of Γ.

Let V be a $Q^c\Gamma$ module with character χ. For $1 \leq i \leq n = \dim_{Q^c}(V)$, let $\lambda^i(\chi)$ be the character of ith exterior product of V. By convention we put $\lambda^0(\chi) = \varepsilon_\Gamma$, the identity character of Γ, and of course $\lambda^i(\chi) = 0$ for $i > n$. Then R_Γ, together with the operations λ^i: $R_\Gamma \to R_\Gamma$, extended from characters to virtual characters by the relation

$$\lambda^i(\chi + \phi) = \sum_{0 \leq j \leq i} \lambda^j(\chi)\, \lambda^{i-j}(\phi)$$

is a λ-ring in the sense of Grothendieck. In particular if $\gamma \in \Gamma$ has eigenvalues $x_1, \ldots x_n$ on V, then for $1 \leq i \leq n$, γ has eigenvalues on the ith exterior product of V,

$$x_{a_1} \ldots x_{a_i}$$

where $1 \leq a_1 < \ldots < a_i \leq n$. In other words $\lambda^i(\chi)(\gamma)$ is the ith symmetric polynomial $S_i(x_1 \ldots x_n)$ in the eigenvalues. By convention $S_0(x_1 \ldots x_n) = 1$, and for $i > 0$ the rth symmetric power sum is

$$P_r(x_1, \ldots, x_n) = \sum_{a=1}^{n} x_a^r .$$

By Newton's formulae

$$n\, S_n = \sum_{k=1}^{n} (-1)^{k-1} P_k \cdot S_{n-k}.$$

Thus, inductively, it can be deduced that the central function

$$\gamma \to \chi(\gamma^i) = P_i(x_1, \ldots, x_n)$$

lies in R_Γ. This virtual character will be written as $\psi_i(\chi)$. The operation ψ_i: $R_\Gamma \to R_\Gamma$ is called the ith Adams operation (or sometimes i^{th} Frobenius operation).

Let p be a prime number and let Γ be a fixed finite group. For $\gamma \in \Gamma$ we may write $\gamma = \gamma' \gamma_p = \gamma_p \gamma'$ uniquely with γ' (resp. γ_p) having order prime to p (resp. p power order).

Suppose $|\Gamma| = p^n m$ with $(p,m) = 1$ and let r be a positive integer r such that

$$r \equiv 0 \bmod (p^n) \qquad r \equiv 1 \bmod (m) .$$

δ_p is defined to be the operation ψ_r. Then clearly for $\chi \in R_\Gamma$, $\gamma \in \Gamma$,

$$\delta_p(\chi)(\gamma) = \chi(\gamma'), \tag{2.2}$$

and $\delta_p^2 = \delta_p$ on R_Γ. These idempotent operations were first introduced by Philippe Cassou-Noguès [CN1].

Let χ be a character of Γ of degree $d = \chi(1)$. Then $\lambda^d(\chi)$ is an abelian character of Γ, and the map $\chi \to \lambda^d(\chi)$ induces a homomorphism

$$\det\colon R_\Gamma \to R_\Gamma^*,$$

which satisfies the relation

$$\det(\chi\phi) = \det(\chi)^{\phi(1)} \det(\phi)^{\chi(1)}$$

for χ, $\phi \in R_\Gamma$.

If Δ is a subgroup of Γ, restriction from Γ to Δ yields a ring homomorphism

$$\mathrm{Res}_\Gamma^\Delta : R_\Gamma \to R_\Delta$$

and induction (i.e. $\otimes_{Q^c\Delta} Q^c\Gamma$) yields a group homomorphism

$$\mathrm{Ind}_\Delta^\Gamma\colon R_\Delta \to R_\Gamma.$$

If $q\colon \Gamma \to \Sigma$ is group epimorphism, then composition with q yields an injective ring homomorphism

$$\mathrm{Inf}_\Sigma^\Gamma\colon R_\Sigma \to R_\Gamma,$$

which will be called inflation.

Let K be a number field, let A be a finite dimensional

semi-simple K-algebra, and let $\mathcal{A}$ be an $\mathcal{O}_K$ order in A. The group of ideles of A, J(A), is the subgroup of elements in ΠA_p^* almost all of whose entries lie in $\mathcal{A}_p^*$. [If p is infinite then $\mathcal{A}_p$ is defined to be A_p]. Note that J(A) does not depend on $\mathcal{A}$, since for any other order $\mathcal{B}$ in A, $\mathcal{B}_p = \mathcal{A}_p$ for almost all primes p. The group of unit ideles of $\mathcal{A}$, U($\mathcal{A}$), is the group $\prod_p \mathcal{A}_p^*$.

For a finite set S of prime numbers define $U_S(\mathcal{A}) = \prod_{p \in S} \mathcal{A}_p^*$, and for a number field F, $U_S(F) = \prod_q \mathcal{O}_{F_q}^*$, where q runs through the primes of F above those primes in S.

In the sequel we take F to be a number field containing K, which is Galois over Q and which is large enough in the sense that the absolutely irreducible classes of representations of A are to be realisable over F.

Given a representation T: $A \to M_n(F)$, twisting by $\omega|_F$ for $\omega \in \Omega_K$, yields a further representation of A. Hence R_A is an Ω_K-module. Now J(F), the ideles of F, are also an Ω_K-module, so that the group $\mathrm{Hom}_{\Omega_K}(R_A, J(F))$ is now defined.

The principal goal of this first chapter is to describe Cl($\mathcal{A}$) as a quotient of this group of homomorphisms. This is Fröhlich's so-called Hom-description of the locally free classgroup. His description is simultaneously well-adapted both to questions of functoriality and to specific calculations. With this end in view, we now introduce various subgroups of homomorphisms.

$\mathrm{Hom}_{\Omega_K}(R_A, F^*)$ will be considered as a subgroup of $\mathrm{Hom}_{\Omega_K}(R_A, J(F))$ via the diagonal map $F^* \hookrightarrow J(F)$.

For $z \in GL_a(\mathcal{A}_p)$ we wish to define an Ω_K homomorphism

$$\mathrm{Det}(z):\ R_A \to F \otimes_K K_p\ .$$

To do this, it suffices by linearity to define Det(z) on classes of $A \otimes_K K^c$ modules, and then check that it is additive and commutes with Ω_K-action. Let T: $A \to M_n(F)$ be a representation of A with class $\chi \in R_A$. Then we extend T to an algebra homomorphism

$$\tilde{T}:\ M_a(A_p) \to M_{na}(F \otimes_K K_p)$$

and define

$$\mathrm{Det}(z)(\chi) = \det(\tilde{T}(z)). \tag{2.3}$$

Additivity is immediate, and commutativity with Ω_K action follows from the equalities

$$(\mathrm{Det}(z)(\chi^{\omega})) = \det(\widetilde{\omega \circ T(z)})$$

$$= \det(\omega(\tilde{T}(z))) \tag{2.4}$$

$$= (\mathrm{Det}(z)(\chi))^{\omega}$$

for $\omega \in \Omega_K$. We remark that Det is related to the map det: $R_\Gamma \to R_\Gamma^*$ by the rule that for $\gamma \in \Gamma$, $\chi \in R_\Gamma$,

$$\mathrm{Det}(\gamma)(\chi) = \det(\chi)(\gamma).$$

If K/L is normal then $\mathrm{Hom}_{\Omega_K}(R_A, J(F))$ may be viewed as an Ω_L-module with action

$$f^{\omega}(\chi) = f(\chi^{\omega^{-1}})^{\omega}$$

for $\omega \in \Omega_L$. Note that in the case $A = \mathcal{O}_K\Gamma$, as in (2.4), it follows that

$$\mathrm{Det}(z)^{\omega} = \mathrm{Det}(z^{\omega}), \tag{2.5}$$

for $z \in \mathcal{O}_{K_p}\Gamma^*$, with Ω_L acting coefficient-wise on $\mathcal{O}_{K_p}\Gamma$.

We shall call the group of homomorphisms Det(z) for $z \in GL_a(A_p)$, $\mathrm{Det}(GL_a(A_p))$. In fact because A_p is semi-local, z can always be multiplied by elementary matrices in $GL_a(A_p)$ to change z into upper triangular form with all but one entry equal to 1 (cf. [B], V, (4.1)). Hence for all $a \geq 1$

$$\mathrm{Det}(GL_a(A_p)) = \mathrm{Det}(A_p^*). \tag{2.6}$$

In exactly the same way, we can define $\mathrm{Det}(GL_a(A_p))$, $\mathrm{Det}(GL_a(A))$ etc. and obtain equalities corresponding to (2.6).

Next the corresponding local construction will be considered together with the relation between the global and local construction. Let $\mathcal{q}$ be a prime of F over a prime $\mathcal{p}$ of K. Then for $z \in A_{\mathcal{p}}^{*}$ we have $\mathrm{Det}(z) \in \mathrm{Hom}_{\Omega_{K_{\mathcal{p}}}}(R_{A_{\mathcal{p}}}, F_{\mathcal{q}}^{*})$, and the natural isomorphisms

$$F \otimes_K K_{\mathcal{p}} \cong \prod_{\mathcal{q}|\mathcal{p}} F_{\mathcal{q}} \qquad A_{p} \cong \prod_{\mathcal{p}|p} A_{\mathcal{p}}$$

induce isomorphisms

$$\mathrm{Det}(A_{p}^{*}) \cong \prod_{\mathcal{p}} \mathrm{Det}(A_{\mathcal{p}}^{*}). \tag{2.7}$$

$$\mathrm{Det}(A_{p}^{*}) \cong \prod_{\mathcal{p}} \mathrm{Det}(A_{\mathcal{p}}^{*}).$$

For an irreducible class $\chi \in R_{A_{\mathcal{p}}}$ and for $z \in A^{*}$, $\mathrm{Det}(z)(\chi)$ is, up to Galois conjugacy, the reduced norm of the image of z in the simple algebra corresponding to χ. We assert that

$$\mathrm{Det}(A_{\mathcal{p}}^{*}) = \begin{cases} \mathrm{Hom}_{\Omega_{K_{\mathcal{p}}}}^{+}(R_{A_{\mathcal{p}}}, F_{\mathcal{q}}^{*}) & \mathcal{p} \text{ real,} \\ \mathrm{Hom}_{\Omega_{K_{\mathcal{p}}}}(R_{A_{\mathcal{p}}}, F_{q}^{*}) & \text{otherwise.} \end{cases} \tag{2.8}$$

where the superscript + on Hom, means that the homomorphisms are to be totally positive on all symplectic representations of $A_{\mathcal{p}}$. This reduces immediately to the case where $A_{\mathcal{p}}$ is simple and the result then follows from the fact that the reduced norm of a local division algebra maps onto the centre, unless $\mathcal{p}$ is real and $A_{\mathcal{p}}$ is a matrix algebra over the quaternions in which case the reduced norm maps onto the positive reals. In the same way, we see that if $\mathcal{p}$ is finite and $A_{\mathcal{p}}$ is a maximal order of $A_{\mathcal{p}}$, then

$$\mathrm{Det}(\mathcal{A}_{\mathcal{p}}^{*}) = \mathrm{Hom}_{\Omega_{K_{\mathcal{p}}}}(R_{A_{\mathcal{p}}}, \mathcal{O}_{F_{\mathcal{q}}}^{*}). \tag{2.9}$$

In the same vein, from the Hasse-Schilling norm theorem (cf. Theorem 7.6 in [SE]), we have

$$\mathrm{Det}(A^{*}) = \mathrm{Hom}_{\Omega_K}^{+}(R_A, F^{*}) \tag{2.10}$$

This section will be concluded by describing the relationship between K_1-groups and the above Det-groups, in certain cases.

Let $\alpha \in K_1(A)$ be represented by $x \in GL_a(A)$. We assert that the map $x \to Det(x)$ induces an isomorphism

$$K_1(A) \cong Det(A^*). \tag{2.11}$$

This follows from the remark after (2.7) together with the fact that the reduced norm yields an isomorphism of $K_1(A)$ into the centre of A, when A is a finitely generated, semi-simple algebra over either a number field or a local field. (Cf. Wang's Theorem in V Theorem 9.7 [B]).

§3. FRÖHLICH'S DESCRIPTION

Let K be a number field and let $\mathcal{p}$ be a finite prime of K. A locally freely presented $A_{\mathcal{p}}$ module is automatically a locally freely presented A module. Moreover a torsion $\mathcal{O}_K$ module is a product of torsion $\mathcal{O}_{K_{\mathcal{p}}}$ modules as $\mathcal{p}$ ranges through the finite primes of K. Thus, by weak approximation,

$$K_oT(A) \cong \prod_{\mathcal{p}} K_oT(A_{\mathcal{p}}). \tag{3.1}$$

Theorem 3.2

Let K be a finite extension of Q_p with $p < \infty$; then there is a natural isomorphism

$$\sigma: K_oT(A) = \frac{Hom_{\Omega_K}(R_A, Q_p^{c*})}{Det(A^*)}$$

Moreover, if M is as in (1.2) with N,P spanning A^n, and $P = N\alpha$ for $\alpha \in GL_n(A)$, then $\sigma([M])$ is represented by the homomorphism $Det(\alpha)$.

By (3.1) we will immediately be able to deduce the corresponding global result:

Theorem 3.3

Let K be a number field, then there is a natural isomorphism

$$\sigma: K_oT(A) \tilde{\to} \frac{Hom^+_{\Omega_K}(R_A, J(F))}{Det(U(A))}.$$

Moreover, if M is as in (1.2) with N,P spanning A^n, and $P_p = N_p.\alpha_p$ for $\alpha_p \in GL_n(A_p)$, then $\sigma[M]$ is represented by that homomorphism which has component $\mathrm{Det}(\alpha_p)$ (resp. 1) at p when p is finite (resp. infinite).

From (2.6) it follows that

Corollary 3.4

$K_0T(A)$ is generated by the classes of torsion modules $A/_b$ where b is a locally free A-ideal.

We now prove Theorem 3.2. Since this is a local situation, the map r in (1.3) is injective, and so

$$K_0T(A) \cong \frac{K_1(A)}{\kappa(K_1(A))} .$$

The required isomorphism follows from (2.11), on observing that $\kappa(K_1(A))$ maps to $\mathrm{Det}(A^*)$.

We now obtain the corresponding description for Cl(A) when K is a number field. From (1.3)

$$\mathrm{Cl}(A) \cong \frac{K_0T(A)}{\theta(K_1(A))} .$$

By the definition of θ it follows that, under the isomorphism of Theorem 3.2, $\theta(K_1(A))$ maps to $\mathrm{Det}(A^*)\,\mathrm{Det}(U(A))/\mathrm{Det}(U(A))$. By (2.10), using weak approximation for the infinite primes:

Theorem 3.5

Let K be a number field, then there is an isomorphism

$$\rho : \mathrm{Cl}(A) \xrightarrow{\sim} \frac{\mathrm{Hom}_{\Omega_K}(R_A, J(F))}{\mathrm{Hom}_{\Omega_K}(R_A, F^*)\,\mathrm{Det}(U(A)).}$$

Moreover, if M is a locally free A module spanning $\overset{n}{\underset{1}{\oplus}} A$, if N is a free A module spanning $\overset{n}{\underset{1}{\oplus}} A$, and if $M_p = N_p\,\alpha_p$ for $\alpha_p \in GL_n(A_p)$; then $\rho[M]$ is represented by the homomorphism with component $\mathrm{Det}(\alpha_p)$ (resp. 1) for p finite (resp. infinite).

Again let K be a number field and we let M be a maximal order of A. Then from (2.9), it follows that

$$\mathrm{Det}(U(\mathcal{M})) = \mathrm{Hom}^+_{\Omega_K}(R_A, U(F))$$

and so Theorem 3.5 gives an isomorphism

$$\mathrm{Cl}(\mathcal{M}) \cong \frac{\mathrm{Hom}_{\Omega_K}(R_A, J(F))}{\mathrm{Hom}_{\Omega_K}(R_A, F^*)\ \mathrm{Hom}^+_{\Omega_K}(R_A, U(F))} \tag{3.6}$$

and so, evaluating on representatives of the Ω_K orbits of the classes of absolutely irreducible representations, we obtain

Theorem 3.7 (Eichler)

Let $A = \Pi A_i$ be the decomposition of A into simple K-algebras and let C_i denote the centre of A_i, then we have an isomorphism

$$\mathrm{Cl}(\mathcal{M}) \cong \prod_i \mathrm{Cl}^+(\mathcal{O}_{C_i})$$

where $\mathrm{Cl}^+(\mathcal{O}_{C_i})$ is the classgroup of $\mathcal{O}_{C_i}$ ideals taken modulo ideals which are principal and totally positive at each place of C_i where A_i is a matrix algebra over a quaternionic division algebra.

(3.8) By the naturality of ρ, it is now trivial that $\mathrm{Cl}(\mathcal{A})$ maps onto $\mathrm{Cl}(\mathcal{M})$, and that the kernel

$$\rho(D(\mathcal{A})) = \frac{\mathrm{Hom}^+_{\Omega_K}(R_A, U(F)).\mathrm{Hom}_{\Omega_K}(R_A, F^*)}{\mathrm{Det}(U(\mathcal{A}))\ .\ \mathrm{Hom}_{\Omega_K}(R_A, F^*)}\ .$$

$$\cong \frac{\mathrm{Hom}^+_{\Omega_K}(R_A, U(F))}{\mathrm{Hom}^+_{\Omega_K}(R_A, \mathcal{O}^*_F)\ \mathrm{Det}(U(\mathcal{A}))} \tag{3.9}$$

(3.10) In particular, we see that $D(\mathcal{A})$ is independent of the particular chosen maximal order $\mathcal{M}$.

Next let S denote a finite set of finite rational primes containing all prime numbers p such that $[\mathcal{M}_p : \mathcal{A}_p] > 1$. Then (3.9) can be rewritten as

$$D(\mathcal{A}) \xrightarrow{\sim} \frac{\mathrm{Hom}_{\Omega_K}(R_A, U_S(F))}{\mathrm{Hom}^+_{\Omega_K}(R_A, \mathcal{O}^*_F)\ .\ \mathrm{Det}(U_S(\mathcal{A}))}\ . \tag{3.11}$$

Reasoning as above, we obtain the analogous result

$$\mathrm{Ker}[K_0T(A) \to K_0T(M)] \underset{\rho}{\overset{\sim}{\to}} \frac{\mathrm{Hom}^+_{\Omega_K}(R_A, U(F))}{\mathrm{Det}(U(A))} .$$

Corollary 3.12

Because $[U(M) : U(A)]$ is finite, and the numerator of the left hand side of the above isomorphism is $\mathrm{Det}(U(M))$

$$|\mathrm{Ker}(K_0T(A) \to K_0T(M)| < \infty.$$

An example We shall now consider the above work when applied to the particular case of Swan modules (cf. [Sw1]). Let Γ be a finite group and let r be a rational integer which is coprime to $|\Gamma|$, the order of Γ. Let $[r,\Sigma_\Gamma]$ denote the right $\mathbb{Z}\Gamma$ module generated over $\mathbb{Z}\Gamma$ by r and $\Sigma_\Gamma = \sum_{\gamma\in\Gamma} \gamma$. Because $|\Gamma|$ is invertible at primes dividing r

$$[r,\Sigma_\Gamma] \otimes_{\mathbb{Z}} \mathbb{Z}_p = x_p \cdot \mathbb{Z}_p \Gamma \tag{3.13}$$

where

$$x_p = \begin{cases} 1 & \text{if } p \nmid r, \\ \dfrac{\Sigma_\Gamma}{|\Gamma|} + (1 - \dfrac{\Sigma_\Gamma}{|\Gamma|})r & \text{if } p \mid r. \end{cases}$$

Thus $[r,\Sigma_\Gamma]$ is a locally free $\mathbb{Z}\Gamma$ module. Let (r,Σ_Γ) denote its class. Since $\mathbb{Z}_p\Gamma$ is a maximal order when $p \nmid |\Gamma|$ (cf. [Re]), S may be taken to be the set of prime divisors of $|\Gamma|$ in the isomorphism (3.11).

Theorem 3.14 (Swan, Ullom)

$(r,\Sigma_\Gamma) \in D(\mathbb{Z}\Gamma)$ and under the isomorphism (3.11), taking S to be the divisors of $|\Gamma|$, this class is represented by the homomorphism

$$\chi \to r^{(\chi,\varepsilon_\Gamma)}$$

for $\chi \in R_\Gamma$. Here (,) is the standard inner product of R_Γ, and ε_Γ is the identity character.

Remark Swan first defined the classes (r, Σ_Γ) and showed that $(r, \Sigma_\Gamma) \in D(\mathbb{Z}\Gamma)$. Ullom gave the above representing homomorphism for (r, Σ_Γ).

Proof From (3.13) (r, Σ_Γ) is represented by $j = \prod_p \mathrm{Det}(x_p)$. Explicitly, for irreducible $\chi \in R_\Gamma$

$$(j(\chi))_p = \begin{cases} r^{\chi(1)} & \chi \neq \varepsilon_\Gamma \text{, and } p \mid r \\ 1 & \text{otherwise} \end{cases},$$

where $\chi(1)$ is the degree of χ. We let $\beta \in U(\mathbb{Z}\Gamma)$ be the idele such that

$$(\beta)_p = \begin{cases} 1 & p \mid r, \\ r & p \nmid r, \end{cases}$$

and we let $g \in \mathrm{Hom}_{\Omega_Q}(R_\Gamma, Q^*)$ be the unique homomorphism such that for each irreducible character χ of Γ

$$g(\chi) = \begin{cases} 1 & \chi = \varepsilon_\Gamma , \\ r^{-\chi(1)} & \chi \neq \varepsilon_\Gamma . \end{cases}$$

Let $f_r = j\,\mathrm{Det}(\beta)g$. Clearly f_r still represents (r, Σ) under (3.5), and, checking all cases, it is seen that for an irreducible character of Γ

$$(f_r(\chi))_p = \begin{cases} 1 & p \mid r \quad \text{or} \quad \chi \neq \varepsilon_\Gamma , \\ r & p \nmid r \text{ and } \chi = \varepsilon_\Gamma . \end{cases}$$

Clearly $f_r \in \mathrm{Hom}^+_{\Omega_Q}(R_\Gamma, U(Q))$, and so the result is shown.

§4. FUNCTORIALITY

For the purpose of this section let K be a number field, let Γ

be a finite group, and let A be the group ring $\mathcal{O}_K\Gamma$. We start by describing the standard homomorphisms, associated to the classgroups we have introduced, with respect to change of group. In the latter part of this section the homomorphisms associated to change in coefficient ring will be considered. It should be emphasised that the results described are all global and that the simplification required to obtain the corresponding local result is omitted.

The results of this section are all taken from the Appendix of [F1].

Induction. Let Δ be a subgroup of Γ, and let M be a locally free $\mathcal{O}_K\Delta$ module. Then $M \otimes_{\mathcal{O}_K\Delta} \mathcal{O}_K\Gamma$ is a locally free $\mathcal{O}_K\Gamma$ module and so one obtains group homomorphisms

$$\mathrm{Ind}_\Delta^\Gamma\colon K_oT(\mathcal{O}_K\Delta) \to K_oT(\mathcal{O}_K\Gamma), \tag{4.1}$$

$$\mathrm{Ind}_\Delta^\Gamma\colon \mathrm{Cl}(\mathcal{O}_K\Delta) \to \mathrm{Cl}(\mathcal{O}_K\Gamma). \tag{4.2}$$

We wish to show that, under the isomorphisms (3.2) and (3.5), these homomorphisms are induced by the homomorphism

$$\mathrm{Ind}_\Delta^\Gamma\colon \mathrm{Hom}_{\Omega_K}(R_\Delta, J(F)) \to \mathrm{Hom}_{\Omega_K}(R_\Gamma, J(F)) \tag{4.3}$$

where for each $f \in \mathrm{Hom}_{\Omega_K}(R_\Delta, J(F))$

$$(\mathrm{Ind}_\Delta^\Gamma f)(\chi) = f(\mathrm{Res}_\Gamma^\Delta \chi)$$

for $\chi \in R_\Gamma$. Since $\delta(K_oT(\mathcal{O}_K\Delta)) = \mathrm{Cl}(\mathcal{O}_K\Delta)$ it suffices to prove for (4.1). Moreover, by Corollary (3.4) and (3.1), it suffices to consider $M = \mathcal{O}_{K_p}\Delta/\alpha\mathcal{O}_{K_p}\Delta$ for $\alpha \in \mathcal{O}_{K_p}\Delta \cap K_p\Delta^*$. Then $[M] \in K_oT(\mathcal{O}_K\Delta)$ is represented by the homomorphism which is $\mathrm{Det}(\alpha)$ at p and 1 at other primes. However

$$M \otimes_{\mathcal{O}_{K_p}\Delta} \mathcal{O}_{K_p}\Gamma \cong \frac{\mathcal{O}_{K_p}\Gamma}{\alpha\mathcal{O}_{K_p}\Gamma},$$

and so the result is immediate. In particular, we remark that from (4.1) we may deduce the already obvious result

$$\mathrm{Ind}_{\Delta}^{\Gamma}(\mathrm{Det}(U(\mathcal{O}_K\Delta))) \subseteq \mathrm{Det}(U(\mathcal{O}_K\Gamma)). \tag{4.4}$$

Passage to quotient. Let $q: \Gamma \to \Sigma$ be a surjective group homomorphism and let $\mathcal{B} = \mathrm{Ker}(\mathcal{O}_K\Gamma \to \mathcal{O}_K\Sigma)$. If M is a locally free $\mathcal{O}_K\Gamma$ module, then $M/_{M\mathcal{B}}$ is a locally free $\mathcal{O}_K\Sigma$ module via q. In this way one obtains group homomorphisms

$$\mathrm{Coinf}_{\Gamma}^{\Sigma}: K_0T(\mathcal{O}_K\Gamma) \to K_0T(\mathcal{O}_K\Sigma), \tag{4.5}$$

$$\mathrm{Coinf}_{\Gamma}^{\Sigma}: \mathrm{Cl}(\mathcal{O}_K\Gamma) \to \mathrm{Cl}(\mathcal{O}_K\Sigma). \tag{4.6}$$

We now show that under the isomorphisms (3.2), (3.5) these homomorphisms are induced by

$$\mathrm{Coinf}_{\Gamma}^{\Sigma}: \mathrm{Hom}_{\Omega_K}(R_\Gamma, J(F)) \to \mathrm{Hom}_{\Omega_K}(R_\Sigma, J(F)), \tag{4.7}$$

where for $f \in \mathrm{Hom}_{\Omega_K}(R_\Gamma, J(F))$ and $\chi \in R_\Sigma$,

$$(\mathrm{Coinf}_{\Gamma}^{\Sigma} f)(\chi) = f(\mathrm{Inf}_{\Sigma}^{\Gamma}\chi).$$

As previous it suffices to consider $M = \mathcal{O}_{K_p}\Gamma/_{\alpha\mathcal{O}_{K_p}\Gamma}$ with $\alpha \in \mathcal{O}_{K_p}\Gamma \cap K_p\Gamma^*$. However,

$$M/_{M\mathcal{B}} = \frac{\mathcal{O}_{K_p}\Sigma}{q(\alpha)\,\mathcal{O}_{K_p}\Sigma}$$

and so the result follows. Observe that from (4.5) we may deduce

$$\mathrm{Coinf}_{\Gamma}^{\Sigma}(\mathrm{Det}(U(\mathcal{O}_K\Gamma)) \subseteq \mathrm{Det}(U(\mathcal{O}_K\Sigma)). \tag{4.8}$$

Restriction Again let Δ be a subgroup of Γ and let $\{\gamma_1 \ldots \gamma_n\}$ be a left transversal of Γ/Δ. If M is a locally free $\mathcal{O}_K\Gamma$ module of rank r, then M is a locally free $\mathcal{O}_K\Delta$ module of rank rn. Thus restriction gives rise to homomorphisms

$$\mathrm{Res}_{\Gamma}^{\Delta}: K_0T(\mathcal{O}_K\Gamma) \to K_0T(\mathcal{O}_K\Delta), \tag{4.9}$$

$$\mathrm{Res}_{\Gamma}^{\Delta}\colon \mathrm{Cl}(\mathcal{O}_K\Gamma) \to \mathrm{Cl}(\mathcal{O}_K\Delta). \tag{4.10}$$

We will show that under the isomorphisms (3.2), (3.5) these homomorphisms are induced by

$$\mathrm{Res}_{\Gamma}^{\Delta}\colon \mathrm{Hom}_{\Omega_K}(R_\Gamma, J(F)) \to \mathrm{Hom}_{\Omega_K}(R_\Delta, J(F)), \tag{4.11}$$

where for $f \in \mathrm{Hom}_{\Omega_K}(R_\Gamma, J(F))$ and $\chi \in R_\Delta$,

$$(\mathrm{Res}_{\Gamma}^{\Delta} f)(\chi) = f(\mathrm{Ind}_{\Delta}^{\Gamma}\ \chi).$$

Again, as previously, we need only show this for (4.9) and indeed there we need only consider a module $M = \mathcal{O}_{K_p}\Gamma/\alpha\mathcal{O}_{K_p}\Gamma$. Thus $[M] \in K_0T(\mathcal{O}_K\Gamma)$ is represented by $\mathrm{Det}(\alpha)$. Let

$$\alpha\ \gamma_i = \Sigma\ \gamma_j\ a_{i,j}$$

with $a_{ij} \in \mathcal{O}_K\Delta$. Since $\{\gamma_i\}$ (resp. $\{\alpha\ \gamma_i\}$) is an $\mathcal{O}_{K_p}\Delta$ basis for $\mathcal{O}_{K_p}\Gamma$ (resp. $\alpha\ \mathcal{O}_{K_p}\Gamma$) we deduce that $[M] \in K_0(\mathcal{O}_K\Gamma)$ is represented by that homomorphism which is $\mathrm{Det}\ (a_{i,j})$ at p and 1 at all other primes. On the other hand, if $T\colon \Delta \to GL_a(F)$ is a representation with character χ, then the representation $T_*\colon \Gamma \to GL_{an}(F)$ which maps γ to $(T(b_{i,j}))$ where $\gamma\ \gamma_i = \gamma_j\ b_{i,j}$ and $b_{i,j} \in \Delta$, has character $\mathrm{Ind}_{\Delta}^{\Gamma}\ \chi$. Thus, we see that

$$\mathrm{Det}(a_{i,j})(\chi) = \mathrm{Det}(\alpha)(\mathrm{Ind}_{\Delta}^{\Gamma}\ \chi)$$

as desired. From (4.9) one deduces the useful result that

$$\mathrm{Res}_{\Gamma}^{\Delta}(\mathrm{Det}(U(\mathcal{O}_K\Gamma)) \subseteq \mathrm{Det}(U(\mathcal{O}_K\Delta)). \tag{4.12}$$

We now turn our attention to change of coefficient ring. Let $F \supset L \supset K$. Then $\otimes_{\mathcal{O}_K} \mathcal{O}_L$ induces the homomorphisms

$$\mathrm{Ind}_K^L : K_0T(\mathcal{O}_K\Gamma) \to K_0T(\mathcal{O}_L\Gamma). \tag{4.13}$$

$$\mathrm{Ind}_K^L : \mathrm{Cl}(\mathcal{O}_K\Gamma) \to \mathrm{Cl}(\mathcal{O}_L\Gamma). \tag{4.14}$$

It is a triviality that, under the isomorphisms of §3, these homomorphisms are induced by the inclusion map

$$\mathrm{Hom}_{\Omega_K}(R_\Gamma, J(F)) \hookrightarrow \mathrm{Hom}_{\Omega_L}(R_\Gamma, J(F)). \tag{4.15}$$

More interestingly, we observe that if M is a locally free $\mathcal{O}_L\Gamma$ module of rank r, then M is a locally free $\mathcal{O}_K\Gamma$ of rank r[L:K]. Thus restriction of scalars induces homomorphisms

$$\mathrm{Res}_L^K : K_0T(\mathcal{O}_L\Gamma) \to K_0T(\mathcal{O}_K\Gamma) \tag{4.16}$$

$$\mathrm{Res}_L^K : \mathrm{Cl}(\mathcal{O}_L\Gamma) \to \mathrm{Cl}(\mathcal{O}_K\Gamma). \tag{4.17}$$

Remark Of course $\mathcal{O}_L\Gamma$ may restrict to a non-free $\mathcal{O}_K\Gamma$-module.

We now show that the above homomorphisms are induced by the norm (or transfer) homomorphism

$$\mathcal{N}_{L/K} : \mathrm{Hom}_{\Omega_L}(R_\Gamma, J(F)) \to \mathrm{Hom}_{\Omega_K}(R_\Gamma, J(F)) \tag{4.18}$$

where for $f \in \mathrm{Hom}_{\Omega_L}(R_\Gamma, J(F))$, $\chi \in R_\Gamma$

$$\mathcal{N}_{L/K}f(\chi) = \prod_\omega f(\chi^{\omega^{-1}})^\omega \tag{4.19}$$

ω running through a right transversal $\Omega_L\backslash\Omega_K$. As always, it suffices to prove for (4.16) in the case $M = \mathcal{O}_{L_p}\Gamma/\alpha\mathcal{O}_{L_p}\Gamma$, with $\alpha \in \mathcal{O}_{L_p}\Gamma \cap L_p\Gamma^*$. Let $\{x_1, \ldots, x_n\}$ be a basis of $\mathcal{O}_{L_p}$ over $\mathcal{O}_{K_p}$, and we suppose

$$\alpha\, x_i = \Sigma\, x_j\, a_{i,j}\,. \tag{4.20}$$

Since $\{\alpha\, x_i\}$ (resp. $\{x_i\}$) is a basis of $\alpha\mathcal{O}_{L_p}\Gamma$ (resp. of $\mathcal{O}_{L_p}\Gamma$) over $\mathcal{O}_{K_p}\Gamma$, the class of $[M] \in K_0T(\mathcal{O}_K\Gamma)$ is represented by the homomorphism which is $\mathrm{Det}(a_{i,j})$ at p, and 1 elsewhere.

From (4.20), it follows that for each $\omega \in \Omega_L\backslash\Omega_K$

$$\alpha^\omega\, x_i^\omega = \Sigma\, x_j^\omega\, a_{i,j}\,.$$

On taking determinants and using the fact that $\det(x_i^\omega)^2 \neq 0$

(being the discriminant of L_p/K_p), it can be seen that

$$\mathrm{Det}(a_{i,j}) = \mathcal{N}_{L/K}(\mathrm{Det}(\alpha)).$$

§5. DUALITY

Let K again be a numberfield and Γ be a finite group. In this section we will describe the effect of duality under the isomorphisms σ, ρ of §3. The results of this section are all due to Frohlich cf. [F1], [F2].

Throughout this section we simultaneously use $\bar{\ }$ to denote the K-linear involution on $K\Gamma$ induced by $\gamma \to \gamma^{-1}$ for $\gamma \in \Gamma$, and, for a virtual character χ of Γ, we write $\bar{\chi}$ for contragredient virtual character.

(5.1) Let M now be a locally freely presented $\mathcal{O}_K\Gamma$ module. The dual of M, written $\hat{M}$, is defined to be $\mathrm{Hom}_{\mathcal{O}_K\Gamma}(M, K\Gamma/\mathcal{O}_K\Gamma)$. Here $\hat{M}$ is viewed as a right $\mathcal{O}_K\Gamma$ module by the rule that for such a homomorphism f

$$f^{\gamma}(m) = \overline{(f(m^{\gamma}))}$$

for $\gamma \in \Gamma$, $m \in M$.

We will show that if $[M] \in K_0T(\mathcal{O}_K\Gamma)$ is represented by a homomorphism f, under (3.3), then $[\hat{M}]$ is represented by the homomorphism $\bar{f}$, where for $\chi \in R_\Gamma$, $\bar{f}(\chi) = f(\bar{\chi})$. By Corollary 3.4, it suffices to prove for $M = \mathcal{O}_{K_p}\Gamma/\alpha.\mathcal{O}_{K_p}\Gamma$ with $\alpha \in \mathcal{O}_{K_p}\Gamma \cap K_p\Gamma^*$. We then have isomorphisms:

$$\hat{M} \cong \mathrm{Hom}_{\mathcal{O}_{K_p}\Gamma}\left(\frac{\mathcal{O}_{K_p}\Gamma}{\alpha\,\mathcal{O}_{K_p}\Gamma}, \frac{K\Gamma}{\bar{\alpha}\,\mathcal{O}_{K_p}\Gamma}\right) \cong \frac{\mathcal{O}_{K_p}\Gamma}{\bar{\alpha}.\mathcal{O}_{K_p}\Gamma}.$$

The latter isomorphism being induced by evaluating a homomorphism on $1 + \alpha\,\mathcal{O}_{K_p}\Gamma$. The result now follows.

(5.2) We now turn our attention to locally free $\mathcal{O}_K\Gamma$-modules. For a locally free $\mathcal{O}_K\Gamma$-module M, we define the dual of M, written $\hat{M}$, to be $\mathrm{Hom}_{\mathcal{O}_K\Gamma}(M, \mathcal{O}_K)$. We shall now show that if $\rho(M)$ is represented by the homomorphism f under (3.5), then $\rho(\hat{M})$ is represented by the homomorphism such that for $\chi \in R_\Gamma$

$$\chi \mapsto f(\bar{\chi})^{-1}. \tag{5.3}$$

With this in mind define π to be the K-linear map $\pi\colon K\Gamma \to K$ such that for $\gamma \in \Gamma$

$$\pi(\gamma) = \begin{cases} 1 & \gamma = 1, \\ 0 & \gamma \neq 1. \end{cases}$$

We define the non-singular Γ pairing

$$< , > : \mathcal{O}_K\Gamma \times \mathcal{O}_K\Gamma \to \mathcal{O}_K$$

by $<x,y> = \pi(x.\bar{y})$ for $x,y \in \mathcal{O}_K\Gamma$.

By corollary (3.4) it suffices to prove (5.3) when M is a locally free $\mathcal{O}_K\Gamma$ ideal. Suppose $M_p = \alpha_p\, \mathcal{O}_{K_p}\Gamma$. For each finite p, < , > induces a non singular pairing

$$< , >_p \; : \alpha_p\, \mathcal{O}_{K_p}\Gamma \times \bar{\alpha}_p^{-1}\, \mathcal{O}_{K_p}\Gamma \to \mathcal{O}_{K_p} \quad .$$

Writing α for the idele with component α_p at $p < \infty$ and 1 when $p = \infty$, we know that M is represented by $\mathrm{Det}(\alpha)$. And by the above

$$\hat{M} \cong K\Gamma \cap \Pi\, \bar{\alpha}_p^{-1}\, \mathcal{O}_{K_p}\Gamma$$

so that $\hat{M}$ is represented by $\mathrm{Det}(\bar{\alpha})^{-1}$, as required.

2. CHARACTER ACTION

Let K be a number field and let Γ be a finite group. In this chapter we shall consider the classgroups $K_0T(\mathcal{O}_K\Gamma)$, $Cl(\mathcal{O}_K\Gamma)$ as modules over the ring of K-characters. Following [Ul], we will do this in such a way as to make the action under the isomorphisms of Chapter 1, §3 immediate. This action, together with Brauer's induction theorems, will provide us with a powerful tool for reducing general questions about modules over groups to questions about modules over Q-p-elementary groups. This is of particular importance since it is for just such groups that the logarithmic techniques of chapter 5 are applicable.

The work of this chapter is due in origin to Swan cf. [Sw2], where he first introduced character action on classgroups. However, the treatment we give is based on S. Ullom's approach as given in [Ul]. In particular, the character action on determinant groups is due to him.

§1. THE SWAN-ULLOM THEOREM

Let K be a number field or a finite extension of Q_p with $p < \infty$. We write $X_\Gamma(K)$ (resp. $R_\Gamma(K)$) for the Grothendieck group of $\mathcal{O}_K\Gamma$-lattices (resp. Grothendieck group of KΓ-modules). Alternatively $R_\Gamma(K)$ is the ring of K-characters of Γ.

$\otimes_{\mathcal{O}_K} K$ induces a surjection

$$j: X_\Gamma(K) \to R_\Gamma(K). \tag{1.1}$$

Let M be a locally free $\mathcal{O}_K\Gamma$-module and let X be an $\mathcal{O}_K\Gamma$-lattice. Then $M \otimes_{\mathcal{O}_K} X$ is a locally free $\mathcal{O}_K\Gamma$-module (with diagonal Γ-action). Swan showed that the class $(M \otimes_{\mathcal{O}_K} X) \in Cl(\mathcal{O}_K\Gamma)$ depends only on the image $j(X)$ in $R_\Gamma(K)$, so that $Cl(\mathcal{O}_K\Gamma)$ becomes an $R_\Gamma(K)$ module.

Remark It is usual to consider only $\mathcal{O}_K\Gamma$ lattices which are $\mathcal{O}_K$-free, in order that the above technique define an action on $K_o(\mathcal{O}_K\Gamma)$. However, because the elements of $Cl(\mathcal{O}_K\Gamma)$ have zero rank, we can avoid this restriction.

We may define an action of $R_\Gamma(K)$ on $Hom_{\Omega_K}(R_\Gamma, J(F))$ (resp. $Hom_{\Omega_K}(R_\Gamma, Q_p^{c*})$) if K is global (resp. local) by the rule

$$(\phi f)(\chi) = f(\bar{\phi}\chi) \tag{1.2}$$

for f such a homomorphism, $\chi \in R_\Gamma$, $\phi \in R_\Gamma(K)$.

When K is a number field we will show that under the isomorphism ρ of 1 (3.5), the Swan action is induced by this character action on homomorphisms. In fact, we shall establish the corresponding result for $K_oT(\mathcal{O}_K\Gamma)$, and then, using the exact sequence 1 (1.3), deduce the result for $Cl(\mathcal{O}_K\Gamma)$. Using the natural isomorphism 1 (3.1) we are reduced to showing:

Theorem 1.3

Let K be a finite extension of Q_p, $p < \infty$. Then $\otimes_{\mathcal{O}_K} X_\Gamma(K)$ induces a module structure on $K_oT(\mathcal{O}_K\Gamma)$ over $R_\Gamma(K)$. More precisely, if X is an $\mathcal{O}_K\Gamma$ lattice with image $\phi \in R_\Gamma(K)$, and if M is a locally freely presented $\mathcal{O}_K\Gamma$ module, whose class in $K_oT(\mathcal{O}_K\Gamma)$ is represented by $f \in Hom_{\Omega_K}(R_\Gamma, Q_p^{c*})$ under 1 (3.2), then $[M \otimes_{\mathcal{O}_K} X]$ is represented by ϕf where

$$\phi f(\chi) = f(\bar{\phi}\chi)$$

for $\chi \in R_\Gamma$.

From the exact sequence 1 (1.3), and the isomorphism 1 (3.1) we can immediately deduce

Theorem 1.4

Let K be a number field. Then $\otimes_{\mathcal{O}_K} X_\Gamma(K)$ induces an $R_\Gamma(K)$ module structure on $Cl(\mathcal{O}_K\Gamma)$ and $K_oT(\mathcal{O}_K\Gamma)$. Furthermore, under the isomorphisms 1 (3.3) and 1 (3.5) this action is induced by the action (1.2) above.

Corollary 1.5

Since $\mathrm{Hom}_{\Omega_K}(R_\Gamma, U(F))$ is $R_\Gamma(K)$ stable, $D(\mathcal{O}_K\Gamma)$ is an $R_\Gamma(K)$ module under the above action.

Proof of Theorem 1.3 Since tensoring with an $\mathcal{O}_K\Gamma$ lattice X is exact in the category of freely presented $\mathcal{O}_K\Gamma$-modules, we see that $\otimes_{\mathcal{O}_K} X$ certainly defines an endomorphism of $K_0T(\mathcal{O}_K\Gamma)$. In order to prove that this endomorphism has the desired effect on a representing homomorphism f, we may, in the usual way, restrict ourselves to the case $M = \mathcal{O}_K\Gamma/\alpha\, \mathcal{O}_K\Gamma$, with $\alpha \in \mathcal{O}_K\Gamma \cap K\Gamma^*$. So, under the isomorphism 1 (3.2), [M] is represented by the homomorphism $\mathrm{Det}(\alpha)$.

Let X be an $\mathcal{O}_K\Gamma$-lattice with basis $x_1, \ldots, x_n$. Let $(c_{i,j}) \in GL_n(\mathcal{O}_K\Gamma)$ be defined by

$$\alpha \otimes x_i = \sum_j (1 \otimes x_j)c_{ij} \tag{1.6}$$

and let $S: \Gamma \to GL_n(\mathcal{O}_K)$ be the representation of Γ afforded by X on the basis $\{x_1, \ldots, x_n\}$. If

$$S(\gamma) = (s_{i,j}(\gamma)) \tag{1.7}$$

then we have:

Lemma 1.8

(a) Let $\alpha = \sum_{\gamma\in\Gamma} \alpha_\gamma \gamma$, with $\alpha_\gamma \in K$. Then $c_{i,j}$ in (1.6) is given by

$$c_{i,j} = \sum_\gamma s_{i,j}(\gamma^{-1})\alpha_\gamma \gamma$$

(b) $\mathrm{Det}(c_{i,j}) = \phi.\mathrm{Det}(\alpha)$, with the notation of (1.2).

Corollary 1.9

For each $\phi \in R_\Gamma(K)$

$$\phi(\mathrm{Det}(\mathcal{O}_K\Gamma^*)) \subseteq \mathrm{Det}(\mathcal{O}_K\Gamma^*).$$

Proof If $\beta \in \mathcal{O}_K\Gamma^*$, then $M = \mathcal{O}_K\Gamma/_{\alpha\beta}\,\mathcal{O}_K\Gamma$. Thus by the above $\phi(\mathrm{Det}(\alpha))$ and

$\phi(\mathrm{Det}(\alpha\beta))$ represent the same class in $K_oT(\mathcal{O}_K\Gamma)$.

Proof of Lemma By the definition of $(S_{i,j}(\gamma))$

$$x_i\ \gamma = \Sigma\ S_{ij}(\gamma)\ x_j$$

and so

$$\alpha \otimes x_i = \sum_\gamma \alpha_\gamma\ \gamma \otimes x_i = \sum_\gamma (1 \otimes x_i\ \gamma^{-1})\gamma\ \alpha_\gamma$$
$$= \sum_j (1 \otimes x_j)\ (\sum_\gamma S_{i,j}(\gamma^{-1})\alpha_\gamma\ \gamma).$$

This establishes (a). In order to show (b), it suffices to observe that if $T: \Gamma \to GL_a(Q_p^c)$ is a representation with character χ, then under the induced algebra homomorphism $T : K\Gamma \to M_a(Q_p^c)$

$$T(C_{j,i}) = (\sum_\gamma S_{j,i}(\gamma^{-1})\ T(\gamma)\ \alpha_\gamma).$$

So, because ${}^tS^{-1} \otimes T$ has character $\bar{\phi}\chi$, the result follows. [Here we use the superscript t to denote the transpose of a matrix].

§2. FROBENIUS MODULE STRUCTURE

In this section we put the above work on character action into the language of Frobenius modules.

Let K be either a number field or a finite extension of Q_p with $p < \infty$. Let M be a contravariant functor from the category of finite groups with monomorphisms, to abelian groups. Then M is said to be a Frobenius module for the Frobenius functor $\Gamma \mapsto R_\Gamma(K)$ if $M(\Gamma)$ is an $R_K(\Gamma)$ module satisfying the following conditions:

(2.1.a) M is covariant with respect to induction

(2.1.b) For $\Delta \subset \Gamma$, $\phi \in R_\Gamma(K)$, $m \in M(\Delta)$,

$$\mathrm{Ind}_\Delta^\Gamma((\mathrm{Res}_\Gamma^\Delta\ \phi).m) = \phi.\mathrm{Ind}_\Delta^\Gamma(m)$$

(2.1.c) For $\Delta \subset \Gamma$, $\phi \in R_\Delta(K)$, $m \in M(\Gamma)$

$$\mathrm{Ind}_\Delta^\Gamma(\phi.\mathrm{Res}_\Gamma^\Delta m) = (\mathrm{Ind}_\Delta^\Gamma\phi).m$$

(2.1.d) For $\Delta \subset \Gamma$, $\phi \in R_\Gamma(K)$, $m \in M(\Gamma)$

$$\mathrm{Res}_\Gamma^\Delta(\phi m) = \mathrm{Res}_\Gamma^\Delta\phi.\mathrm{Res}_\Gamma^\Delta m.$$

<u>Theorem 2.2</u>

The functors $\Gamma \mapsto \mathrm{Cl}(\mathcal{O}_K\Gamma)$, $\Gamma \mapsto K_oT(\mathcal{O}_K\Gamma)$ are Frobenius modules for $R_K(\Gamma)$.

<u>Proof</u> (2.1.a) is immediate. The proofs of the remaining parts are all very similar. We therefore only prove (2.1.c). Let $f \in \mathrm{Hom}_{\Omega_K}(R_\Gamma, J(F))$ (resp. $\mathrm{Hom}_{\Omega_K}(R_\Gamma, Q_p^{c*})$) represent m under the isomorphism of 1 §3 if K is a number field (resp. a finite extension of Q_p). For $\chi \in R_\Gamma$ by Frobenius reciprocity:

$$\begin{aligned}(\mathrm{Ind}_\Delta^\Gamma\phi)f(\chi) &= f(\mathrm{Ind}_\Delta^\Gamma\bar{\phi}.\chi)\\ &= f(\mathrm{Ind}_\Delta^\Gamma(\bar{\phi}.\mathrm{Res}_\Gamma^\Delta\chi))\\ &= (\mathrm{Res}_\Gamma^\Delta f)(\bar{\phi}.\mathrm{Res}_\Gamma^\Delta\chi)\\ &= \mathrm{Ind}_\Delta^\Gamma(\phi.\mathrm{Res}_\Gamma^\Delta f).\end{aligned}$$

3. SWAN MODULES

Let Γ be a finite group and let r be an integer which is coprime to the order of Γ. The swan module $[r, \Sigma_\Gamma]$ was defined in chapter 1, §3. In this chapter we give a systematic account of the basic properties of Swan modules. This will be followed by the description of Swan classes for all p-groups (although this result will be proved in chapter 7).

In the final section of this chapter we use Swan modules to prove the self-duality theorem for rings of integers of tame extensions cf. [T4].

The main reason for studying Swan modules is that they provide a canonical sub-group of $D(\mathbb{Z}\Gamma)$ which is relatively easy to handle. In particular Swan classes have excellent functorial properties, they can be calculated explicitly in very many cases, and furthermore, for a number of interesting "small groups" they actually generate all of $D(\mathbb{Z}\Gamma)$.

§1. BASIC PROPERTIES

Let S be the set of prime divisors of the group order $|\Gamma|$. From 1 (3.11) we have an isomorphism

$$D(\mathbb{Z}\Gamma) \cong \frac{\mathrm{Hom}_{\Omega_Q}(R_\Gamma, U_S(F))}{\mathrm{Hom}^+_{\Omega_Q}(R_\Gamma, \mathcal{O}_F^*)\,\mathrm{Det}(U_S(\mathbb{Z}\Gamma))} . \tag{1.1}$$

From chapter 1, §3 we know that the class (r, Σ_Γ) lies in $D(\mathbb{Z}\Gamma)$ and is represented under (1.1) by the homomorphism f_r

$$f_r(\chi) = r^{(\chi, \varepsilon_\Gamma)} \quad \text{for } \chi \in R_\Gamma .$$

Clearly, if r' is also an integer coprime to $|\Gamma|$ then,

$f_r f_{r'} = f_{rr'}$, so that

$$([r,\Sigma_\Gamma] \oplus [r',\Sigma_\Gamma]) = (rr',\Sigma_\Gamma). \tag{1.2}$$

We let $\mathcal{B} = \mathbb{Z}\Gamma/_{\mathbb{Z}\,\Sigma_\Gamma}$. The augmentation map $\varepsilon_\Gamma\colon \mathbb{Z}\Gamma \to \mathbb{Z}$, induces a ring homomorphism

$$\tilde{\varepsilon} : \mathcal{B} \to \mathbb{Z}/_{|\Gamma|\,\mathbb{Z}} \quad .$$

Proposition 1.3

Let r, r' be as above. If $\tilde{\varepsilon}(r) = \tilde{\varepsilon}(w\, r')$ for some $w \in \mathcal{B}^*$, then $(r,\Sigma_\Gamma) = (r',\Sigma_\Gamma)$.

From (1.2) and the above proposition we see that the map $r \mapsto (r,\Sigma_\Gamma)$ induces a homomorphism

$$\partial_\Gamma\colon (\mathbb{Z}/_{|\Gamma|\,\mathbb{Z}})^* \to D(\mathbb{Z}\Gamma) \tag{1.4}$$

and that $\tilde{\varepsilon}(\mathcal{B}^*) \subseteq \mathrm{Ker}(\partial_\Gamma)$. Following Ullom's notation we write $T(\mathbb{Z}\Gamma)$ for $\mathrm{Im}(\partial_\Gamma)$.

Remark 1. Using the Fibre diagram

$$\begin{array}{ccc} \mathbb{Z}\Gamma & \longrightarrow & \mathcal{B} \\ {\scriptstyle\varepsilon}\downarrow & & \downarrow \\ \mathbb{Z} & \longrightarrow & \mathbb{Z}/_{|\Gamma|\mathbb{Z}} \end{array}$$

and the Mayer-Vietoris sequence of Reiner and Ullom it can easily be shown that $\mathrm{Ker}\,\partial_\Gamma$ is the image of $K_1(\mathcal{B})$ in $K_1(\mathbb{Z}/_{|\Gamma|\mathbb{Z}}) = (\mathbb{Z}/_{|\Gamma|\mathbb{Z}})^*$ under the homomorphism induced by $\tilde{\varepsilon}$.

Remark 2. In [Sw1] Swan shows $[r,\Sigma_\Gamma] \cong [r',\Sigma_\Gamma]$ if, and only if, $\tilde{\varepsilon}(r^{-1}r') \in \tilde{\varepsilon}(\mathcal{B}^*)$.

Proof of Proposition $[r,\Sigma_\Gamma]$ is isomorphic to the $\mathbb{Z}\Gamma$-module with generators and relations

$$\langle u,v \mid v\gamma = v \text{ for } \gamma \in \Gamma, \quad u\Sigma_\Gamma = rv\rangle.$$

We have the corresponding presentation for $[r',\Sigma_\Gamma]$ on generators u',v'. Assume now that $\tilde{\varepsilon}(r) = \tilde{\varepsilon}(w)\ \tilde{\varepsilon}(r')$ for $w \in B^*$, and choose some $\alpha \in \mathbb{Z}\Gamma$ such that α maps to w under the natural homomorphism $\mathbb{Z}\Gamma \to B$. The homomorphism induced by

$$u \mapsto \alpha u' \qquad\qquad v \mapsto v'$$

is then seen to be an isomorphism by repeating the process from $\langle u',v'\rangle$ to $\langle u,v\rangle$, with w^{-1} in place of w.

Corollary 1.5

If Γ is a cyclic group, then $T(\mathbb{Z}\,\Gamma) = \{1\}$.

Proof It suffices to show that given a class $m \bmod |\Gamma|$ in $(\mathbb{Z}/_{|\Gamma|\mathbb{Z}})^*$, with $m > 0$, there exists $y_m \in B^*$ with $\tilde{\varepsilon}(y_m) \equiv m \bmod |\Gamma|$. If γ is a generator of Γ, then set

$$x_m = \sum_{i=0}^{m-1} \gamma^i.$$

Clearly, $\tilde{\varepsilon}(x_m + \mathbb{Z}\,\Sigma_\Gamma) = m \bmod |\Gamma|$, and if m' is a positive integer such that $mm' \equiv 1 \bmod |\Gamma|$, then

$$x_m.x_{m'} = 1 + mm'\ \Sigma_\Gamma$$

so that $x_m + \mathbb{Z}\,\Sigma_\Gamma$ is indeed a unit of B.

Next we consider the functorial properties of the Swan group $T(\mathbb{Z}\,\Gamma)$ (cf. [U4]).

If $q: \Gamma \to \Lambda$ is a surjective homomorphism of groups, then it is clear that q maps (r,Σ_Γ) to (r,Σ_Λ) and thereby induces a surjection

$$q_T: T(\mathbb{Z}\,\Gamma) \to T(\mathbb{Z}\,\Lambda). \tag{1.6}$$

If $\Delta \subset \Gamma$, we observe that from 1 §4 the class $\mathrm{Res}^\Delta_\Gamma(r,\Sigma_\Gamma)$ is represented by the homomorphism

$$\chi \mapsto r^{(\mathrm{Ind}_{\Delta}^{\Gamma} \chi, \varepsilon_{\Gamma})}$$

but $(\mathrm{Ind}_{\Delta}^{\Gamma} \chi, \varepsilon_{\Gamma}) = (\chi, \varepsilon_{\Delta})$, and so (r, Σ_{Γ}) restricts to (r, Σ_{Δ}). Thus we obtain the exceedingly useful result that restriction induces a surjection

$$\mathrm{Res}_{\Gamma}^{\Delta}: T(\mathbb{Z}\,\Gamma) \to T(\mathbb{Z}\,\Delta). \tag{1.7}$$

We shall now consider the action of $R_{\Gamma}(Q)$ on $T(\mathbb{Z}\,\Gamma)$. The Artin ideal of $R_{\Gamma}(Q)$, A_{Γ}, is defined by

$$A_{\Gamma} = \{\chi \in R_{\Gamma}(Q) \,|\, \chi = \sum_i \mathrm{Ind}_{\Delta_i}^{\Gamma} \phi_i, \quad \phi_i \in R_{\Delta_i}(Q), \quad \Delta_i \text{ cyclic}\}.$$

A_{Γ} is an ideal by Frobenius reciprocity. We call the characteristic of the finite ring $R_{\Gamma}(Q)/_{A_{\Gamma}}$, the Artin exponent of Γ. The Artin exponent is studied in detail in [L].

Proposition 1.8 (Ullom)
The exponent of $T(\Gamma)$ divides the Artin exponent of Γ.

Proof For $t \in T(\mathbb{Z}\,\Gamma)$, $\chi = \mathrm{Ind}_{\Delta}^{\Gamma}\phi \in A_{\Gamma}$, we see that from 2 (2.1.c)

$$(\mathrm{Ind}_{\Delta}^{\Gamma}\phi)t = \mathrm{Ind}_{\Gamma}^{\Gamma}(\phi.\mathrm{Res}_{\Gamma}^{\Delta}t) = 1$$

by corollary 1.5 above.

Remark In the above we have shown that $A_{\Gamma}.T(\mathbb{Z}\Gamma) = 1$. In the case when Γ is a p-group and p is odd, Robert Oliver has recently shown that A_{Γ} is the exact annihilator of $T(\mathbb{Z}\,\Gamma)$ over $R_Q(\Gamma)$ (cf. [02]). This result is of particular interest in the light of Theorem 2.5 in the next section.

§2. THE p-GROUP CASE

We will now describe the group $T(\mathbb{Z}\,\Gamma)$ when Γ is a p-group. First of all we need to introduce three important families of 2-groups.

Δ_n, the dihedral group of order 2^n, $n \geq 2$, has generators and relations

$$<\alpha, \beta \mid \alpha^{2^{n-1}} = \beta^2 = 1, \quad \beta\alpha\beta = \alpha^{-1}>. \tag{2.1}$$

H_n, the quaternion group of order 2^n, $n \geq 3$, has generators and relations

$$<\sigma, \tau \mid \sigma^{2^{n-2}} = \tau^2, \ \tau^4 = 1, \ \tau^{-1}\sigma\,\tau = \sigma^{-1}>. \tag{2.2}$$

SD_n, the semi-dihedral group of order 2^n, $n \geq 4$, has generators and relations

$$<\xi, \eta \mid \xi^{2^{n-1}} = \eta^2 = 1, \ \ \eta\xi\eta = \xi^{-1+2^{n-2}}> \tag{2.3}$$

(2.4) We shall say that a p-group is exceptional if, and only if, $p = 2$ and the group is either dihedral, quaternion, or semi-dihedral.

<u>Theorem 2.5</u>

(a) Let Γ be a finite p-group. If Γ is non cyclic and is not exceptional then $T(\mathbb{Z}\,\Gamma)$ has order $|\Gamma|p^{-1}$ (resp. $\frac{1}{4}\,|\Gamma|$) if $p \neq 2$ (resp. $p = 2$).

(b) If $p = 2$ and Γ is semi-dihedral or quaternion (resp. dihedral) then $T(\mathbb{Z}\,\Gamma)$ has order two (resp. order 1).

(c) In all cases $T(\mathbb{Z}\,\Gamma)$ is cyclic with $(1 + p, \Sigma_\Gamma)$ as a generator.

<u>Remark 1</u>. Part (a) was shown in [T1]. Part (b), for the dihedral and quaternion cases, is proved in [FKW], while the semi-dihedral result is due to Endo.

<u>Remark 2</u>. In [U2] it is shown that if Γ is a p-group with $|\Gamma| > 2$, then the exponent of $D(\mathbb{Z}\,\Gamma)$ divides $p^{-1}|\Gamma|$ if p is odd (resp. divides $\frac{1}{4}\,|\Gamma|$ if $p = 2$). Thus for non-cyclic, non-exceptional p-groups we see that $(1 + p, \Sigma_\Gamma)$ attains this bound in exponent.

Part (c) follows immediately from (1.3) and (1.4), since $1 + p$ generates the p-part of $(\mathbb{Z}/_{|\Gamma|}\mathbb{Z})^*/<\pm 1>$ in all cases and because (2.1) shows that $D(\mathbb{Z}\,\Gamma)$ is necessarily a p-group.

We will prove part (a) in chapter 7 §1, and part (b) in chapter 7, §2. In chapter 7 §2 it will also be shown that:

Theorem 2.6.

If Γ is an exceptional 2-group, then $T(\mathbb{Z}\Gamma) = D(\mathbb{Z}\Gamma)$.

Theorem 2.7.

If p is a fixed prime and Γ_n any sequence of p-groups with $\lim_{n\to\infty} |\Gamma_n| \to \infty$ and with only a finite number of the Γ_n exceptional when $p = 2$ then

$$\lim_{n\to\infty} |D(\mathbb{Z}\Gamma_n)| \to \infty.$$

Proof By Theorem 2.5 it suffices to show that if C_{p^n} is the cyclic group of order p^n, then

$$\lim_{n\to\infty} |D(\mathbb{Z}C_{p^n})| \to \infty$$

and this is shown in [F2], [G] and [KM]. The interested reader may like to prove this by constructing images of $D(\mathbb{Z}C_{p^n})$ by the technique used in chapter 7, §1.

Reiner and Ullom have shown that if $\{\Gamma_n\}$ is a sequence of abelian groups of composite order, then $\lim_{n\to\infty} |D(\mathbb{Z}\Gamma_n)| \to \infty$. However, this is not true in the non-abelian case, as is demonstrated by the case of dihedral groups D_{2p} of order 2p with p a prime, since $|D(\mathbb{Z}D_{2p})| = 1$ (cf. chapter 4, §3).

§3. SELF-DUALITY OF RINGS OF INTEGERS

As an application we will now use the above work to obtain the self-duality theorem for rings of integers of tame extensions.

We recall that an extension of number fields $L/_K$ is said to be tame if $Tr_{L/K}$, the trace from L to K, maps $\mathcal{O}_L$ onto $\mathcal{O}_K$. By a standard argument (cf. 14.4 [Sel] for instance), it is known that if L/K is tame and Galois with $\Gamma = Gal(L/K)$, then $\mathcal{O}_L$ is a projective $\mathcal{O}_K\Gamma$-module. So, by restriction and by 1 (1.1), $\mathcal{O}_L$ defines a class $(\mathcal{O}_L) \in Cl(\mathbb{Z}\Gamma)$.

The inverse different is defined by

$$D^{-1}_{L/K} = \{x \in L \mid Tr_{L/K}(xy) \in \mathcal{O}_K \quad \forall\, y \in \mathcal{O}_L\}.$$

$Tr_{L/K}$ therefore induces an isomorphism

$$D_{L/K}^{-1} \cong \mathrm{Hom}_{\mathcal{O}_K\Gamma}(\mathcal{O}_L,\mathcal{O}_K).$$

Thus, with the terminology of chapter 1 §5, $(D_{L/K}^{-1}) = (\hat{\mathcal{O}}_L)$.

Theorem 3.1.

Let L/K be a tame Galois extension of number fields with $\Gamma = \mathrm{Gal}(L/K)$. Then $(\mathcal{O}_L) = (D_{L/K}^{-1})$ in $\mathrm{Cl}(\mathbb{Z}\Gamma)$; that is to say $\mathcal{O}_L$ is a stably self-dual $\mathbb{Z}\Gamma$-module.

Remark This result was first shown in [T4] under the slightly more restrictive, so-called "domestic" hypothesis, that the prime divisors of [L:K] are non-ramified in L/K. The extension to the general tame case together with the elegant idea of using the torsion module $D_{L/K}^{-1}/\mathcal{O}_L$, in place of representing homomorphisms is due to Steve Chase.

Proof By the above $M = D_{L/K}^{-1}/\mathcal{O}_K$ is a locally freely presented $\mathbb{Z}\Gamma$-module. From the exact sequence 1 (1.3), it is clear that it is sufficient to show $\delta([M]) = 1$ in $\mathrm{Cl}(\mathbb{Z}\Gamma)$.

Indeed, by the isomorphism 1 (3.1) together with the localisation properties of differents, it suffices to consider

$$M_p = \frac{D_{L/K}^{-1} \otimes_{\mathcal{O}_K} \mathcal{O}_{K_p}}{\mathcal{O}_L \otimes_{\mathcal{O}_K} \mathcal{O}_{K_p}}$$

for each prime p of $\mathcal{O}_K$ and show $\delta([M_p]) = 1$ in $K_0T(\mathbb{Z}\Gamma)$.

Let P denote a prime ideal of L above p. Let Δ (resp. I) denote the decomposition group (resp. the inertia group) of P in L/K. Then Δ (resp. I) identifies with the Galois group of L_P/K_p (resp. of L_P/N_P, where N_P is the maximal non-ramified extension of K_p in L_P). We put

$$M_P = D_{L_P/K_p}^{-1}/\mathcal{O}_{L_P}.$$

Because Γ transitively permutes the primes of L above p,

$$M_p \cong M_P \otimes_{\mathcal{O}_{K_p}\Delta} \mathcal{O}_{K_p}\Gamma.$$

$e = |I|$ is the ramification index of P in L/K and since L/K is tame

$$D_{L_P/K_p} = P^{e-1}\,\mathcal{O}_{L_P}.$$

Thus we obtain an isomorphism of $\mathcal{O}_{K_p}\Delta$ modules

$$M_P \cong \frac{p^{-1}P\mathcal{O}_{L_P}}{\mathcal{O}_{L_P}} \cong \frac{P\mathcal{O}_{L_P}}{p\mathcal{O}_{L_P}}$$

By a simple counting argument we deduce that the natural injection

$$\frac{P\mathcal{O}_{L_P}}{p\mathcal{O}_{L_P}} \hookrightarrow \frac{\mathcal{O}_{L_P}}{p\mathcal{O}_{L_P} + \mathcal{O}_{N_P}}$$

is onto, whence

$$M_P \cong \frac{\mathcal{O}_{L_P}}{p\mathcal{O}_{L_P} + \mathcal{O}_{N_P}} \quad .$$

Taking into account the fact that $\mathcal{O}_{L_P}$ is $\mathcal{O}_{K_p}\Delta$ free, we obtain an isomorphism of Δ-modules

$$M_P \cong \frac{\mathcal{O}_{K_p}\Delta}{(\mathcal{O}_{K_p}\Sigma_I + p\mathcal{O}_{K_p}I) \otimes_{\mathcal{O}_{K_p}I} \mathcal{O}_{K_p}\Delta}$$

where, as usual, $\Sigma_I = \sum_{i \in I} i$. Thus from the above we obtain an isomorphism of $\mathbb{Z}\Delta$ modules

$$M_P \cong \bigoplus_1^f \frac{\mathbb{Z}I}{[p,\Sigma_I]} \otimes_{\mathbb{Z}I} \mathbb{Z}\Delta$$

where f denotes the residue class extension degree of p over p. However, by tameness, $(p, |I|) = 1$, and so $[p,\Sigma_I]$ is a Swan module. Furthermore we know I to be cyclic (being a quotient of $\mathcal{O}^*_{N_P}$ mod $p\,\mathcal{O}_{N_P}$ cf. IV, §2 in [Se2]). Hence by Corollary 1.5, $[p,\Sigma_I]$ is a free $\mathbb{Z}I$ module, □

4. REDUCTION THEORY

Following his description of classgroups, Fröhlich introduced a family of quotient groups (the groups $E_\ell(\Gamma)$) of the kernel group $D(\mathbb{Z}\Gamma)$, by using determinantal congruences, derived from Brauer's decomposition maps for character rings. In fact, such congruences, phrased in terms of reduced norms, had been previously used by C.T.C. Wall in §2 of [W1].

In section 1 we give Fröhlich's congruences and construct the E-groups. In section 2 we apply these techniques to obtain various explicit results for classgroups. In section 3 we prove the Galovich-Reiner-Ullom theorem describing $D(\mathbb{Z}\Gamma)$ when Γ is a so-called pq group (or generalised dihedral group). Although the style of our proof is new, the main ideas are drawn from [GRU] (see also [F3]).

§1. THE MOD ℓ CONGRUENCES

We let $\mathcal{O}_p^c$ (resp. $\mathcal{P}^c$) denote the ring of integers of $\mathbb{Q}_p^c$ (resp. the maximal ideal in the ring of integers of $\mathbb{Q}_p^c$). We shall identify $\mathcal{O}_p^c/\mathcal{P}^c$ with $\mathbb{F}_p^c$, a separable closure of the field with p elements. We write $R_\Gamma(\mathbb{F}_p^c)$ for the Grothendieck group of $\mathbb{F}_p^c\Gamma$-modules.

Let $\chi \in R_{\Gamma,p}$ be the character of a given $\mathbb{Q}_p^c\Gamma$ module V. We choose a Γ-stable $\mathcal{O}_p^c$ lattice X in V, which spans V. Then $\bar{X} = X/X\mathcal{P}^c$ is an $\mathbb{F}_p^c\Gamma$ module whose class in $R_\Gamma(\mathbb{F}_p^c)$ depends only on the character of V. We write $d_{p,\Gamma}(\chi)$ for this class, and so obtain Brauer's decomposition homomorphism (cf. [Se1] §15).

$$d_{p,\Gamma} : R_{\Gamma,p} \to R_\Gamma(\mathbb{F}_p^c) . \tag{1.1}$$

An element $\gamma \in \Gamma$ is said to be p regular if the order of γ is prime to p.

It is shown in Corollary 2 to Theorem 42 in [Se1] that

$$\text{Ker } d_{p,\Gamma} = \{\chi \in R_{\Gamma,p} \mid \chi(\gamma) = 0 \quad \forall\ p \text{ regular } \gamma \in \Gamma\} \tag{1.2}$$

More generally if $h: Q^c \hookrightarrow Q_p^c$, then composition with h yields a decomposition map which is also denoted $d_{p,\Gamma}$

$$d_{p,\Gamma} : R_\Gamma \to R_\Gamma(\mathbb{F}_p^c)$$

Clearly $d_{p,\Gamma}$ depends on the embedding h; however, by (1.2), Ker $d_{p,\Gamma}$ is independent of h.

Lemma 1.3

Let K be a finite extension of Q_p and let $\chi \in \text{Ker } d_{p,\Gamma}$. Then for $z \in \mathcal{O}_K\Gamma^*$

$$\text{Det}(z)(\chi) \equiv 1 \text{ mod } P^c.$$

Proof We write r for the reduction map r: $\mathcal{O}_K\Gamma \to \mathbb{F}_p^c\Gamma$. Because taking determinants commutes with the ring homomorphism r

$$r(\text{Det}(z)(\chi)) = \text{Det}(r(z))\ (d_{p,\Gamma}(\chi)) = 1.$$

The corresponding semi-local version is:

Lemma 1.4

Let $K \subset F$ be a number field and let $\mathfrak{r}_p$ be the radical of $p\mathcal{O}_F$ in $\mathcal{O}_F$. For $\chi \in \text{Ker } d_{p,\Gamma}$ and for $z \in \mathcal{O}_{K_p}\Gamma^*$,

$$\text{Det}(z)(\chi) \equiv 1 \text{ mod } \mathfrak{r}_p.$$

We shall now use these congruences to construct certain quotient groups of $D(\mathbb{Z}\Gamma)$. We recall that from 1 (3.11)

$$D(\mathbb{Z}\Gamma) \cong \frac{\text{Hom}_{\Omega_Q}(R_\Gamma, U_S(F))}{\text{Hom}^+_{\Omega_Q}(R_\Gamma, \mathcal{O}_F^*)\ \text{Det}(U_S(\mathbb{Z}\Gamma))}, \tag{1.5}$$

where S is the set of primes dividing the order of Γ.

We define $V_p = (\mathcal{O}_p/\mathfrak{r}_p)^*$. The composite of

$$\mathrm{Ker}\, d_{p,\Gamma} \hookrightarrow R_\Gamma \to U_S(F) \to \mathcal{O}^*_{F_p} \to V_p \qquad (1.6)$$

induces a homomorphism

$$h'_p: \mathrm{Hom}_{\Omega_Q}(R_\Gamma, U_S(F)) \to \mathrm{Hom}_{\Omega_Q}(\mathrm{Ker}\, d_{p,\Gamma}, V_p) \qquad (1.7)$$

Lemma 1.8

h'_p is a surjection.

Proof V_p is an Ω_K-direct factor of $\mathcal{O}^*_{F_p}$ (being the complement to the pro-Sylow p-subgroup of $\mathcal{O}^*_{F_p}$). Thus it will suffice to show $\mathrm{Ker}\, d_{p,\Gamma}$ is an Ω_K direct factor of R_Γ.

Let δ_p be the Adams operation, which is an idempotent for R_Γ, defined in 1 (2.2). It is immediate from (1.2) that $\delta_p(\mathrm{Ker}\, d_{p,\Gamma}) = 0$ and that $(1 - \delta_p) R_\Gamma \subseteq \mathrm{Ker}\, d_{p,\Gamma}$. Thus $(1 - \delta_p) R_\Gamma = \mathrm{Ker}\, d_{p,\Gamma}$ and so the result follows from the Ω_K isomorphism

$$R_\Gamma = (1 - \delta_p)R_\Gamma \oplus \delta_p R_\Gamma. \qquad \square$$

By Lemma 1.4, $\mathrm{Det}(U_S(\mathbb{Z}\Gamma)) \subseteq \mathrm{Ker}(h'_p)$ and so, if we set

$$E_p(\Gamma) = \frac{\mathrm{Hom}_{\Omega_Q}(\mathrm{Ker}\, d_{p,\Gamma}, V_p)}{h'_p(\mathrm{Hom}^+_{\Omega_Q}(R_\Gamma, \mathcal{O}^*_F))},$$

then h'_p induces a surjection

$$h_p: D(\mathbb{Z}\Gamma) \to E_p(\Gamma).$$

Remark Let M be a maximal order of $Q\Gamma$ which contains $\mathbb{Z}\Gamma$. Ph. Cassou-Noguès has shown that if $\Lambda^{(p)}$ is that order with the property that $\Lambda^{(p)}_q$ is

(1) M_q, at $q \neq p$

(2) the order generated by $\mathbb{Z}_p\Gamma$ together with the Jacobson radical of M_p, at $q = p$; then $E_p(\Gamma)$ is naturally isomorphic to $D(\Lambda^{(p)})$.

§2. SOME GENERAL RESULTS

Our first aim is to show:

Theorem 2.1. (Reiner-Ullom).
Let Γ be a finite p-group, then $D(\mathbb{Z}\Gamma)$ is a p-group.

Remark If Γ is an abelian p-group, then Fröhlich has calculated the order of $D(\mathbb{Z}\Gamma)$ (resp. the order of $D(\mathbb{Z}\Gamma)^-$, the subgroup of stably self-dual locally free $\mathbb{Z}\Gamma$ modules) when $p = 2,3$ (resp. $p > 3$) cf. [F5] and [F6].

Proof Under the isomorphism (1.5) we may now take $S = \{p\}$. Observe further, that we may choose F to be a cyclotomic field obtained by adjoining the $|\Gamma|$th roots of unity to Q.

If $p = 2$, then $\mathrm{Hom}_{\Omega_Q}(R_\Gamma, U_2(F))$ is a finitely generated $\mathbb{Z}_2$-module, and since $\mathrm{Det}(\mathbb{Z}_2\Gamma^*)$ has finite index, the result is a triviality in this case.

So now suppose $p > 2$. Since Γ has no irreducible symplectic characters

$$\mathrm{Hom}^+_{\Omega_Q}(R_\Gamma, \mathcal{O}^*_F) = \mathrm{Hom}_{\Omega_Q}(R_\Gamma, \mathcal{O}^*_F).$$

Let $f \in \mathrm{Hom}_{\Omega_Q}(R_\Gamma, U_p(F))$ and again let $\mathfrak{r}_p$ denote the Jacobson radical of $p\mathcal{O}_{F_p}$ in $\mathcal{O}_{F_p}$.

Let $\varepsilon_\Gamma = \chi_1, \chi_2, \ldots \chi_n$ represent the distinct Ω_Q orbits of the irreducible characters of Γ. By Galois action, $f(\chi_i)$ lies in $(\mathbb{Z}(\chi_i) \otimes_{\mathbb{Z}} \mathbb{Z}_p)^*$ where $\mathbb{Z}(\chi_i)$ is the ring of integers of $Q(\chi_i)$, the field generated over Q by the values of χ_i on Γ. Now p is totally ramified in F and so

$$f(\chi_i) \equiv n_i \mod \mathfrak{r}_p$$

for some positive integer coprime to p. Thus $f(\chi_i^{\omega}) \equiv n_i \bmod \mathfrak{r}_p$, for $\omega \in \Omega_Q$.

The following elementary result is well-known:

Lemma 2.2
Let $p \neq 2$, let n be a positive integer coprime to p and let ζ_p

be a primitive p^{th} root of unity. Then $\xi_n = (\zeta_p^n - \zeta_p^{-n})(\zeta_p - \zeta_p^{-1})^{-1}$ is a unit in the ring of integers of $Q(\zeta_p + \zeta_p^{-1})$ and

$$\xi_n \equiv n \mod (1 - \zeta_p).$$

Now $n_1 \in \mathbb{Z}_p^* \subset \mathbb{Z}_p\Gamma^*$, and so $f\,\mathrm{Det}(n_1)^{-1}$ represents the same class as f. We define $u \in \mathrm{Hom}_{\Omega_Q}(R_\Gamma, \mathcal{O}_F^*)$ by

$$u(\chi_i) = \xi_{n_i}^{-1} \xi_{n_1}.$$

Again we know that $f' = f\, u\, \mathrm{Det}(n_1)^{-1}$ represents the same class and moreover

$$f'(\chi_i^\omega) \equiv 1 \mod \mathfrak{p}_p$$

for $\omega \in \Omega_Q$, $1 \leqslant i \leqslant n$. Thus f' lies in the pro-p-group $\mathrm{Hom}_{\Omega_Q}(R_\Gamma, 1 + \mathfrak{p}_p)$ and the result follows because $\mathrm{Det}(\mathbb{Z}_p\Gamma^*)$ has finite index in $\mathrm{Hom}_{\Omega_Q}(R_\Gamma, U_p(F))$ (cf. 1 (3.12)).

<u>Corollary 2.3</u> (Reiner-Rim)
Let C be a group of order p, then

$$D(\mathbb{Z}C) = \{1\}.$$

<u>Proof</u> From (1.5), we know that $D(\mathbb{Z}C)$ is a quotient of the group

$$X = \frac{\mathrm{Hom}_{\Omega_Q}(R_C, U_p(F))}{\mathrm{Det}(\mathbb{Z}_p C^*)}.$$

Let χ be a non-trivial abelian character of C. By Galois action, it is clear that $f \in \mathrm{Hom}_{\Omega_Q}(R_C, U_p(F))$ is determined by the two values $f(\varepsilon_C)$, $f(\chi)$ in $\mathbb{Z}_p^*$ (resp. $\mathbb{Z}_p[\zeta_p]^*$). Let γ generate C. We see immediately that the sub-group $\mathrm{Det}(1 + (1-\gamma)\mathbb{Z}_p C)$ (resp. $\mathrm{Det}(1 + \mathbb{Z}\Sigma_C)$) is that subgroup of homomorphisms g such that

$$g(\varepsilon_C) = 1 \quad g(\chi) \equiv 1 \mod (1 - \zeta_p)$$

resp. $$g(\varepsilon_C) \equiv 1 \mod (p) \; g(\chi) = 1.$$

In consequence X, and whence $D(\mathbb{Z}C)$, is annihilated by $p - 1$. However, by the above theorem $D(\mathbb{Z}C)$ is a p-group.

§3. GENERALISED DIHEDRAL GROUPS

Let p be a prime number, and let $q \mid p - 1$. We write Δ for the generalised dihedral group which has presentation:

$$\langle \gamma, \delta \mid \gamma^p = \delta^q = 1, \quad \delta^{-1}\gamma\delta = \gamma^{\alpha}\rangle \tag{3.1}$$

where $\alpha \in \mathbb{F}_p^*$ has order q.

The main result of this section is

Theorem 3.2 (Galovich-Reiner-Ullom)

Let $q' = q/_{(q,2)}$, and let Δ^{ab} denote Δ made abelian; then

$$\mathrm{Ker}(D(\mathbb{Z}\Delta) \to D(\mathbb{Z}\Delta^{ab})) = T(\mathbb{Z}\Delta)$$

is a cyclic group of order q'.

From Corollary 2.3 we deduce

Corollary 3.3

Suppose, further, that q itself is a prime number, then $D(\mathbb{Z}\Delta)$ is a cyclic group of order q'.

Remark The result in the case $q = 2$ is due to J. Martinet.

Let ζ_p be a primitive p^{th} root of unity of Q^c. Once and for all we fix an embedding $Q(\zeta_p) \hookrightarrow Q_p^c$, and so view elements of $Q(\zeta_p)$ as elements of Q_p^c.

Let L be the unique subfield of $Q(\zeta_p)$ with $(Q(\zeta_p):L) = q$. We let $g \in \Sigma = \mathrm{Gal}(Q(\zeta_p)/L)$ be that automorphism with the property that $\zeta_p^g = \zeta_p^{\alpha}$. We shall identify Σ and $\langle\delta\rangle$ $(\subset\Delta)$ via the map $g \mapsto \delta$. Let $\mathbb{Z}[\zeta_p]_0\Sigma$ be the twisted group ring (with trivial cocycle) which has $\mathbb{Z}[\zeta_p]\Sigma$ as underlying $\mathbb{Z}$-module, and multiplication given by:

$$(z\sigma).(z'\sigma') = zz'^{\sigma^{-1}}\sigma\sigma'$$

for $z, z' \in \mathbb{Z}[\zeta_p]$, $\sigma, \sigma' \in \Sigma$. Clearly $\mathcal{O}_L = \mathbb{Z}[\zeta_p]^{\Sigma}$ is the centre of

$\mathbb{Z}[\zeta_p]_o\Sigma$.

Lemma 3.4

$D(\mathbb{Z}[\zeta_p]_o\Sigma) = \{1\}$.

Proof We start by showing that every $\mathbb{Z}$-free $\mathbb{Z}[\zeta_p]_o\Sigma$ module P is projective, so that $\mathbb{Z}[\zeta_p]_o\Sigma$ is hereditary. Our proof is taken from 14.4 in [Sel]. Let P_o denote the underlying $\mathbb{Z}[\zeta_p]$ module and let π denote the homomorphism induced by the identity

$$P_o \otimes_{\mathbb{Z}[\zeta_p]} \mathbb{Z}[\zeta_p]_o\Sigma \xrightarrow{\pi} P$$

We split π by demonstrating the existence of $u \in \mathrm{End}\,(P_o)$ such that $\sum_{\sigma\in\Sigma} u(\sigma^{-1}p) \otimes \sigma \overset{\pi}{\mapsto} p$, for all $p \in P$. However, $\mathbb{Q}(\zeta_p)/L$ is tame so there exists $v \in \mathbb{Z}[\zeta_p]$ with trace 1 in $\mathcal{O}_L$. We now view $v \in \mathrm{End}\,(P_o)$ and set $u = v$. The lemma now follows by a general result of Jacobinski (cf. [J]) on observing that $\mathbb{Q}(\zeta_p)_o\Sigma$ satisfies the Eichler condition.

Remark For the calculation of kernel groups of twisted group rings see [Wi].

The abelian character $\chi: \langle\gamma\rangle \to \langle\zeta_p\rangle$ given by $\chi(\gamma) = \zeta_p$, induces a surjective homomorphism of algebras

$$\chi_* : \mathbb{Z}\Delta \to \mathbb{Z}[\zeta_p]_o\Sigma. \tag{3.5}$$

We must now introduce a number of specific characters of Δ. Let $\rho = \mathrm{Ind}_{\langle\gamma\rangle}^{\Delta}\, \varepsilon_{\langle\gamma\rangle}$, that is to say ρ is the regular character of Δ^{ab} inflated to Δ. Let $\eta = \mathrm{Ind}_{\langle\gamma\rangle}^{\Delta}\, \chi$, $\xi = \eta - \rho$. Then $\chi - \varepsilon_{\langle\gamma\rangle} \in \mathrm{Ker}\, d_{p,\langle\gamma\rangle}$, and, since $\mathrm{Ker}\, d_p$ is inductive (cf. (1.2)), we know $\xi \in \mathrm{Ker}\, d_{p,\Gamma}$. Thus evaluation on ξ induces a surjection

$$\xi_* : D(\mathbb{Z}\Delta) \to E_p(\Delta) \to \frac{(\mathcal{O}_L \bmod \mathfrak{p})^*}{r(\mathcal{O}_L^*)} \tag{3.6}$$

where $\mathfrak{p}$ is the prime ideal of L over p and we write $r(\mathcal{O}_L^*)$ for the image of $\mathcal{O}_L^*$ under reduction mod $\mathfrak{p}$.

In the sequel we write

$$G = \frac{(\mathcal{O}_L \bmod p)^*}{r(\mathcal{O}_L^{\ *})} .$$

The proof of Theorem 3.2 will go in two parts:

(A) Show $|D(\mathbb{Z}\,\Delta)| \leq |G|$.

(B) Show G is a cyclic group of order q´.

Proof of (A) Let R_1 (resp. R_2) denote the Grothendieck group of $Q\Delta^{ab} \otimes Q^c$ (resp. $Q(\zeta_p)\circ\Sigma \otimes Q^c$) modules. We shall identify R_2 with $\mathbb{Z}\,\Omega_Q.\eta$, and we abbreviate $\mathrm{Hom}_{\Omega_Q}(R_i, \mathcal{O}_F^*)$ to H_i, and $\mathrm{Hom}_{\Omega_Q}(R_\Delta, \mathcal{O}_F^*)$ to H. Note that for Δ, the superscript + on Hom_{Ω_Q} is superfluous, because Δ has no irreducible symplectic characters.

The quotient map $\Delta \to \Delta^{ab}$, together with inclusion, induces a commutative diagram

$$\begin{array}{ccccccccc}
1 \to & H_2 \prod_t \mathrm{Det}(\mathbb{Z}[\zeta_p]_t \circ\Sigma^*) & \to & H \prod_t \mathrm{Det}(\mathbb{Z}[\zeta_p]_t \circ\Sigma^* \times \mathbb{Z}_t\,\Delta^{ab^*}) & \to & H_1 \prod_t \mathrm{Det}(\mathbb{Z}_t\,\Delta^{ab^*}) & \to 1 \\
 & \uparrow{\scriptstyle j} & & \uparrow{\scriptstyle i} & & \| & \\
1 \longrightarrow & K & \longrightarrow & H \prod_t \mathrm{Det}(\mathbb{Z}_t\,\Delta^*) & \longrightarrow & H_1 \prod_t \mathrm{Det}(\mathbb{Z}_t\,\Delta^{ab^*}) & \to 1
\end{array}$$

(3.7)

where K is the kernel of the lower right hand map; so that both rows are exact. Here t ranges through the primes dividing $|\Delta|$ and, as always, the subscript t denotes $\otimes_{\mathbb{Z}} \mathbb{Z}_t$.

Now $(1 - \gamma)\mathbb{Z}\,\Delta = \mathrm{Ker}(\mathbb{Z}\Delta \to \mathbb{Z}\,\Delta^{ab})$ and $(1 - \zeta_p)$ is invertible at all primes different from p. Hence

$$K \supseteq H_2[\prod_{t\neq p} \mathrm{Det}(\mathbb{Z}[\zeta_p]_t \circ\Sigma^* \times \mathrm{Det}(1 + (1 - \zeta_p)\mathbb{Z}_p[\zeta_p]\circ\Sigma)] \qquad (3.8)$$

We denote the right hand term by K´.

Lemma 3.9

For $t \neq p$

$$\mathrm{Det}(\mathbb{Z}[\zeta_p]_t \circ \Sigma^*) = \mathrm{Hom}_{\Omega_Q}(R_2, U_t(F)).$$

Proof We are required to show that

$$\mathrm{Det}(\mathbb{Z}[\zeta_p]_t \circ \Sigma^*)(\eta) = \mathcal{O}^*_{L_t}.$$

This follows from the following two results:

(i) For $z \in \mathbb{Z}[\zeta_p]_t$,

$$\mathrm{Det}(z)(\eta) = N_{Q(\zeta_p)_t/L_t}(z).$$

(ii) $N_{Q(\zeta_p)_t/L_t}$ maps $\mathbb{Z}[\zeta_p]^*_t$ onto $\mathcal{O}^*_{L_t}$, since t is non-ramified in $Q(\zeta_p)/L$.

Corollary 3.10

Let π be the natural projection

$$\pi: \mathrm{Hom}_{\Omega_Q}(R_2, \prod_t U_t(F)) \to \mathrm{Hom}_{\Omega_Q}(R_2, \prod_{t \neq p} U_t(F)).$$

Then, $$\pi(H_2 \prod_t \mathrm{Det}(\mathbb{Z}[\zeta_p]_t \circ \Sigma^*)) = \prod_{t\neq p} \mathrm{Det}(\mathbb{Z}[\zeta_p]_t \circ \Sigma^*)$$

and so π induces an isomorphism:

$$\frac{H_2 \prod_t \mathrm{Det}(\mathbb{Z}[\zeta_p]_t \circ \Sigma^*)}{K'} \cong \frac{H_2\ \mathrm{Det}(\mathbb{Z}_p[\zeta_p] \circ \Sigma^*)}{H_2\ \mathrm{Det}(1 + (1-\zeta_p)\mathbb{Z}_p[\zeta_p] \circ \Sigma)}$$

Remark We are viewing H_2 as homomorphisms with values in $\prod_t U_t(F)$ on the right and with values in $U_p(F)$ on the left. This is quite standard as, in all our local calculations, we assume global elements to be embedded by the appropriate diagonal map.

From 1 (3.11), we know that Coker(i) in (3.7) is isomorphic to $\mathrm{Ker}(D(\mathbb{Z}\Delta) \to D(\mathbb{Z}[\zeta_p] \circ \Sigma \times \mathbb{Z}\Delta^{ab}))$. By the Snake lemma Coker(i) $\cong$ Coker(j), and from (3.8) and (3.10) we see that Coker(j) is a quotient of

$$\frac{\mathcal{O}_L^*.\mathrm{Det}(\mathbb{Z}_p[\zeta_p]o\Sigma^*)(\eta)}{\mathcal{O}_L^*\ .\ \mathrm{Det}(1 + (1 - \zeta_p)\ \mathbb{Z}_p[\zeta_p]o\Sigma)(\eta).} \tag{3.11}$$

Hence in order to show (A) it now suffices to show

$$\mathrm{Det}(1 + (1 - \zeta_p)\ \mathbb{Z}_p[\zeta_p]o\Sigma)(\eta) \supseteq 1 + p\mathcal{O}_{L_p}.$$

Now for $z \in \mathbb{Z}_p[\zeta_p]^*$,

$$\mathrm{Det}(1 + (1 - \zeta_p)z)(\eta) = N_{Q(\zeta_p)_p/L_p}(1 + (1 - \zeta_p)z)$$

while, on the other hand, because $Q(\zeta_p)/L$ is tame,

$$N_{Q(\zeta_p)_p/L_p}(1 + (1 - \zeta_p)\mathbb{Z}_p[\zeta_p]) = 1 + p\mathcal{O}_{L_p}.$$

<u>Proof of (B)</u> Let $[L:Q] = m$ and let $Q(\zeta_p)^+ = Q(\zeta_p + \zeta_p^{-1})$.
We identify the residue class ring $\mathbb{Z}[\zeta_p]$ mod $(1 - \zeta_p)$ with $\mathbb{F}_p$.

Since L/Q is totally ramified at p, it follows that the norm $N_{L/Q}$ acts on $\mathcal{O}_L$ mod p by raising to the mth power. Clearly $N_{L/Q}(\mathcal{O}_L^*) \subseteq \langle\pm 1\rangle$; so that $N_{L/Q}$ induces a homomorphism from G onto $(\mathbb{F}_p^*)^m(\pm 1)/(\pm 1)$, a cyclic group of order q'. Therefore, in order to establish (B), it suffices to show that $r(\mathcal{O}_L^*)$ contains the subgroup $(\mathbb{F}_p^*)^q$ (resp. $(\mathbb{F}_p^*)^{q/2}$) if q is odd (resp. is even). This now follows from Lemma 2.2 and the observation that

$$N_{Q(\zeta_p)/L}(\xi_a) \equiv a^q \quad \mathrm{mod}\ p,$$

and that furthermore when q is even

$$N_{Q(\zeta_p)^+/L}(\xi_a) \equiv a^{q/2} \quad \mathrm{mod}\ p.$$

(B) is now shown.

It only remains to show that

$$T(\mathbb{Z}\Delta) = \mathrm{Ker}(D(\mathbb{Z}\Delta) \to D(\mathbb{Z}\Delta^{ab}))\ .$$

The containment $\subseteq$ is immediate from 3 (1.5), since Δ^{ab} is cyclic. To show the other containment, we need only observe that for a positive integer r prime to pq

$$\xi_*((r,\Sigma_\Delta)) = r^{(\varepsilon_\Gamma,\xi)} \quad r(\mathcal{O}_L^*) \bmod p$$

from (3.6) and 1 (3.14). Since $(\varepsilon_\Gamma,\xi) = -1$, it is immediate that $\xi_*(T(\mathbb{Z}\Delta)) = G$, as we require.

5. TORSION DETERMINANTS

The aim of this chapter is to prove a fundamental result of C.T.C. Wall on the torsion of determinants of group rings. This result is absolutely vital to our study because it complements the group logarithm introduced in the next chapter. This is because the group logarithm describes determinants of local group rings modulo torsion.

The results of this chapter are taken from [W1]. Let Γ be a finite p-group with $|\Gamma| = p^n$. Let K be a finite extension of $\mathbb{Q}_p$. We denote the group of p-power roots of unity in K by μ_K, and, for the sake of brevity, we shall write $\mathcal{O}$ in place of $\mathcal{O}_K$.

<u>Theorem 1.1</u> (C.T.C. Wall)

Let $x \in \mathcal{O}\Gamma^\times$ have the property that Det(x) has finite, p-power order. Then there exist $\eta \in \mu_K$, $\gamma \in \Gamma$ such that

$$\mathrm{Det}(x) = \mathrm{Det}(\eta\gamma).$$

We remark that for any $\gamma \in \Gamma$, $\eta = \mathrm{Det}(\eta\gamma)(\varepsilon_\Gamma)$, thus we are at liberty to enlarge K, and so henceforth we shall suppose that K contains the p^nth roots of unity.

<u>Step 1</u> We start by establishing the theorem when Γ is abelian.

<u>Lemma 1.2</u>

If Δ is any abelian group, then the map $y \mapsto \mathrm{Det}(y)$ induces an isomorphism

$$\mathcal{O}\Delta^* \xrightarrow{\sim} \mathrm{Det}(\mathcal{O}\Delta^*)$$

<u>Proof</u> Let $z = \sum z_\delta \delta \in \mathcal{O}\Delta^*$ and suppose $\mathrm{Det}(z) = 1$. Then, summing over abelian characters χ of Δ,

$$|\Delta|\, z_\delta = \sum_\chi \chi(z.\delta^{-1})$$

$$= \sum \chi(\delta^{-1}).\mathrm{Det}(z)(\chi)$$

$$= \begin{cases} |\Delta| & \delta = 1, \\ 0 & \text{otherwise.} \end{cases}$$ □

<u>Lemma 1.3</u>

Theorem 1.1 holds when Γ is abelian.

<u>Proof</u> Let $v\colon K^* \to \mathbb{Z}^+$ be the normalised valuation with $v(p) = 1$, and let ε_Γ denote the identity character of Γ. By multiplying by $\mathrm{Det}(x)(\varepsilon_\Gamma)^{-1}$ $(\in \mu_K)$ we may assume, without loss of generality, that $\mathrm{Det}(x)(\varepsilon_\Gamma) = 1$.

Let $\gamma \in \Gamma$ have order p, and set $\bar{\Gamma} = \Gamma/_{\langle\gamma\rangle}$. Given an object associated to Γ, adjunction of $^-$ is to denote the corresponding object for $\bar{\Gamma}$. By induction on the group order we may suppose $\mathrm{Det}(\bar{x}) = 1$, i.e. by Lemma 1.2, $x \in 1 + (1 - \gamma)\,\mathcal{O}\Gamma$.

Let ϕ denote any abelian character of Γ with $\phi(\gamma) \neq 1$. Then $\phi(1 - x) \in (1 - \zeta_p)\mathcal{O}$, where ζ_p denotes a primitive p^{th} root of unity. Now, by hypothesis $\phi(1 - x) = 1 - \zeta$ for some $\zeta \in \mu_K$, and so, by considering valuations, we deduce immediately that $\phi(x) = \phi(\gamma)^a$ for some integer a. Thus, writing $x\gamma^{-a}$ for x, we may now assume that $\phi(x) = 1$. Note that $(1 - x)$ still belongs to $(1 - \gamma)\,\mathcal{O}\Gamma$,

$$(1 - x) = (1 - \gamma)y \tag{1.4}$$

with $y \in \mathcal{O}\Gamma$. However $\phi(1 - x) = 0$, so that $\phi(y) = 0$, whence

$$y \in \sum_{\delta\in\Gamma} \mathcal{O}(\delta - \phi(\delta)) \tag{1.5}$$

the free $\mathcal{O}$ module on $\{\delta - \phi(\delta)\}$.

We now assume that $x \neq 1$ for a contradiction. We may then choose an abelian character ψ of Γ such that $\psi(x) \neq 1$. By (1.4)

$$v(\psi(1 - x)) = v(\psi(1 - \gamma)) + v(y).$$

However, from (1.5) we know $v(y) > 0$, while necessarily $v(\psi(1 - \gamma)) \geq v(1 - \zeta_p)$. Therefore, because $v(\psi(1 - x)) = v(1 - \zeta)$ for some $\zeta \in \mu_K$, we see that $\psi(x) = 1$. □

Step 2 Next we suppose that $\Delta \subset \Gamma$, with $(\Gamma:\Delta) = p$ (thus Δ is normal in Γ) and we suppose in addition that Δ is abelian. By Lemma 1.2 there is a unique element $N(x) \in \mathcal{O}\Delta^*$ such that

$$\mathrm{Res}_\Gamma^\Delta(\mathrm{Det}(x)) = \mathrm{Det}(N(x)). \tag{1.6}$$

Let $\gamma \in \Gamma\setminus\Delta$, and let $x = \sum_{\theta\in\Gamma} x_\theta\, \theta$. Conjugation by γ induces an automorphism α of Γ

$$\gamma^{-1}\, \theta\, \gamma = \theta^\alpha.$$

We let T denote the free $\mathcal{O}$ module on the elements $\sum_{i=1}^{p} \theta^{\alpha^i}$, for $\theta \in \Gamma$, and let $t\colon \Gamma \to \Delta$ denote the transfer homomorphism.

Lemma 1.7

$$N(x) \equiv \sum x_\theta^p\, t(\theta) \bmod T$$

Proof By the definition of $\mathrm{Res}_\Gamma^\Delta$ (cf. 1 §4) we know that if $x = \sum_{j=0}^{p-1} \gamma^i\, a_j$ with $a_j \in \mathcal{O}\Delta$, then

$$\mathrm{Res}_\Gamma^\Delta(\mathrm{Det}(x)) = \mathrm{Det}(x_{i,j})$$

where for $0 \leq i,j < p$,

$$x_{i,j} = \begin{cases} \gamma^p\, a_{j-i}^{\alpha^i} & \text{if} \quad j < i \\ a_{j-i}^{\alpha^i} & \text{if} \quad j \geq i \end{cases}$$

Here, where necessary, we view the suffices on a and x as elements of $\mathbb{Z}/_{p\mathbb{Z}}$. We now observe that

$$x_{i,j}^{\alpha} = \begin{cases} x_{i+1,j+1} & i,j < p-1 \\ \gamma^{p} x_{0,j+1} & j < i = p-1 \\ \gamma^{-p} x_{i+1,0} & i < j = p-1 \\ x_{0,0} & i = j = p-1 \end{cases}$$

Now $\mathrm{Det}(x_{i,j}) = \sum_{\sigma} y(\sigma)$, where

$$y(\sigma) = \pi(\sigma) \prod_{o}^{p-1} x_{i,\sigma(i)}$$

σ runs through all permutations of $\{0,1,\ldots p-1\}$ and $\pi(\sigma)$ denotes the parity of σ. We let τ denote the cycle $(0,1,\ldots,p-1)$. Then,

$$\begin{aligned} y(\sigma)^{\alpha} &= \pi(\sigma) \prod_{o}^{p-1} x_{i+1,\sigma(i)+1} \\ &= \pi(\sigma) \prod_{o}^{p-1} x_{\tau(i),\tau\sigma(i)} \\ &= y(\tau\sigma\tau^{-1}). \end{aligned}$$

Thus, if $\sigma^{\tau} \neq \sigma$ we obtain a contribution of $\sum_{i=0}^{p-1} y(\sigma^{\tau^{i}})$ towards $\mathrm{Det}(x_{i,j})$, and this lies in T. So, because $\sigma^{\tau} = \sigma$ if, and only if $\sigma \in \langle\tau\rangle$, we have shown

$$\mathrm{Det}(x_{i,j}) \equiv \sum_{i=0}^{p-1} \gamma^{ip} \prod_{k=0}^{p-1} a_{i}^{\alpha^{k}} \mod T .$$

To conclude our proof we need only show that for $r,s \in \mathcal{O}\Delta$

$$\prod_{o}^{p-1} (r+s)^{\alpha^{k}} \equiv \prod_{o}^{p-1} r^{\alpha^{k}} + \prod_{o}^{p-1} s^{\alpha^{k}} \mod T, \tag{1.8}$$

and then observe that $t(\gamma) = \gamma^{p}$, while for $\delta \in \Delta$, $t(\delta) = \prod_{o}^{p-1} \delta^{\alpha^{k}}$. With this in mind we show:

Lemma 1.9

Let $\{\cdot, X_i, Y_j, \cdot\}$ be 2p commuting algebraically independent indeterminates, with $i, j \in \mathbb{Z}/_{p\mathbb{Z}}$. We let $\langle\alpha\rangle$ act on the $\{\cdot, X_i, Y_j, \cdot\}$ by $X_i^\alpha = X_{i+1}$, $Y_j^\alpha = Y_{j+1}$. Then

$$\prod_{k=0}^{p-1} (X_0 + Y_0)^{\alpha^k} = \prod_{0}^{p-1} X_0^{\alpha^k} + \prod_{0}^{p-1} Y_0^{\alpha^k} + \sum_{0}^{p-1} Z^{\alpha^k}$$

for some $Z \in \mathbb{Z}[\cdot, X_i, Y_j, \cdot]$

Proof The left hand side is α fixed and each mixed monomial in the product expansion occurs once only. Thus all its $\langle\alpha\rangle$ conjugates must occur once and once only. □

(1.8) now follows immediately from the above lemma by substitution.

Proposition 1.10

For Γ, Δ, x as above, $\mathrm{Res}_\Gamma^\Delta(\mathrm{Det}(x)) = 1$.

Proof Clearly $\mathrm{Res}_\Gamma^\Delta(\mathrm{Det}(x)) = \mathrm{Det}(N(x))$ is of p-power order, because Det(x) is. Furthermore, using Step 1 and multiplying by suitable elements in Γ and μ_K, we may assume that x has image 1 under the quotient map

$$q: \mathcal{O}\Gamma \to \mathcal{O}\Gamma/_{[\Gamma,\Gamma]} \, .$$

We now observe that an irreducible character χ of Γ is either abelian, in which case $\mathrm{Det}(x)(\chi) = 1$, or $\chi = \mathrm{Ind}_\Delta^\Gamma \phi$ for some abelian character ϕ of Δ, and so $\mathrm{Det}(x)(\chi) = \mathrm{Det}(N(x))(\phi)$. Thus it will suffice to show $N(x) = 1$.

Because Δ is abelian, we know by step 1 that $N(x) = r\delta_0$ for some $r \in \mu_K$, $\delta_0 \in \Delta$. To show that $r = 1$, we observe that

$$\begin{aligned} r = \mathrm{Det}(N(x))(\varepsilon_\Delta) &= (\mathrm{Res}_\Gamma^\Delta \, \mathrm{Det}(x))(\varepsilon_\Delta) \\ &= \mathrm{Det}(x)(\mathrm{Ind}_\Delta^\Gamma \, \varepsilon_\Delta) \\ &= 1 \end{aligned}$$

since $q(x) = 1$ and $\Delta \supset [\Gamma,\Gamma]$.

Let $x = \Sigma\, x_\gamma \gamma$. Because $q(x) = 1$

$$\sum_{\theta\in\gamma[\Gamma,\Gamma]} x_\theta = \begin{cases} 1 & \gamma \in [\Gamma,\Gamma], \\ 0 & \text{otherwise} \end{cases} .$$

On raising to the pth power, we see that

$$\sum_{\theta\in\gamma[\Gamma,\Gamma]} x_\theta^p \equiv \begin{cases} 1 & \gamma \in [\Gamma,\Gamma], \\ & \bmod(p) \\ 0 & \gamma \notin [\Gamma,\Gamma]. \end{cases}$$

By Lemma 1.7 it now follows that

$$N(x) \equiv \Sigma\, x_\theta^p\, t(\theta) \equiv 1 \quad \bmod\ (T,p)$$

and so $\delta_o - 1 \in (T,p)$. However, any coefficient of 1 in (T,p) is necessarily divisible by p, and so $\delta_o = 1$, as was required. □

<u>Step 3</u> We now use the above special cases to establish the general result. Again we assume that x has image 1 under the quotient map $q\colon \mathcal{O}\Gamma \to \mathcal{O}\Gamma^{ab}$. We must show that for any irreducible character χ of Γ, $\mathrm{Det}(x)(\chi) = 1$. By standard character theory we can find $\Delta \subset \Gamma$ with $(\Gamma:\Delta) = p$ such that there is a character ϕ of Δ which induces χ on Γ.

Next we pick $x' \in \mathcal{O}\Delta^*$ with the property that

$$\mathrm{Res}_\Gamma^\Delta\, \mathrm{Det}(x) = \mathrm{Det}(x')$$

and we observe that $\mathrm{Det}(x')$ again has p-power order.

We now have a commutative diagram with natural maps:

$$\begin{array}{ccccc} \mathrm{Det}(\mathcal{O}\Gamma^*) & \xrightarrow{r} & \mathrm{Det}(\mathcal{O}\Gamma/_{[\Delta,\Delta]}^{\;*}) & \xrightarrow{q'} & \mathrm{Det}(\mathcal{O}\Gamma^{ab*}) \\ \downarrow{\scriptstyle \mathrm{Res}} & & \downarrow{\scriptstyle \mathrm{Res}} & & \\ \mathrm{Det}(\mathcal{O}\Delta^*) & \xrightarrow{r'} & \mathrm{Det}(\mathcal{O}\Delta^{ab*}) & & \end{array}$$

and $1 = q(\mathrm{Det}(x)) = q'r\,\mathrm{Det}(x)$. Applying Step 2 to the group $\Gamma/_{[\Delta,\Delta]}$, we deduce that

$$r(\mathrm{Det}(x)) = 1$$

and hence $r' \,\mathrm{Res}_{\Gamma}^{\Delta}(\mathrm{Det}(x)) = 1$, i.e. $r'(\mathrm{Det}(x')) = 1$. Thus, arguing by induction on the group order, $\mathrm{Det}(x') = 1$, and therefore

$$\begin{aligned} 1 = \mathrm{Det}(x')(\phi) &= (\mathrm{Res}_{\Gamma}^{\Delta}\, \mathrm{Det}(x))(\phi) \\ &= \mathrm{Det}(x)(\chi) \ . \end{aligned}$$

□

6. THE GROUP LOGARITHM

In the Hom-descriptions of Grothendieck groups of group rings given in chapter 1, we obtain factors of the type $\mathrm{Det}(\mathbb{Z}_p\Gamma^*)$ in the denominator. Thus, in order to be able to see how close a representing homomorphism lies to the denominator, we have to have a much better understanding of $\mathrm{Det}(\mathbb{Z}_p\Gamma^*)$. This is achieved by means of the group logarithm.

In the first section we begin by stating the main results concerning the group logarithm, for the case of p-groups. We then prove a number of results, but we defer the proof of the integrality property of the logarithm until section 2. In section 3 we extend our techniques to deal with Q- p-elementary groups. The relevance of this tool for classgroups of group rings will become more clear in subsequent chapters, where it is used to prove several major results.

Nearly all the results of this chapter are taken from [T1] - though it should be mentioned that the origins of this work are to be found in [T2].

§1. THE MAIN RESULTS

Let Γ be a finite p-group, let K be a finite non-ramified extension of Q_p, and let $\Delta = \mathrm{Gal}(K/Q_p)$. As in 1 (2.5), $\mathrm{Det}(\mathcal{O}_K\Gamma^*)$ is viewed as a Δ-module by the rule

$$\mathrm{Det}(z)^\delta = \mathrm{Det}(z^\delta)$$

for $z \in \mathcal{O}_K\Gamma^*$, $\delta \in \Delta$, with δ acting coefficientwise on z (cf. 1 (2.5)).

Let $K(|\Gamma|)$ be the field obtained by adjoining the $|\Gamma|^{th}$ roots of unity in Q_p^c to K. Then $K(|\Gamma|)$ is the composite of K and the totally ramified extension $Q_p(|\Gamma|)$. We let f be the Frobenius automorphism of the extension $K(|\Gamma|)/Q_p(|\Gamma|)$.

Let $\mathfrak{r}_K$ denote the Jacobson radical of $\mathcal{O}_K\Gamma$. Explicitly $\mathfrak{r}_K$ is generated over $\mathcal{O}_K$ by p and $1 - \gamma$ for $\gamma \in \Gamma$. The image of $\mathfrak{r}_K$ under reduction mod p is the Jacobson radical of the Artinian ring $\mathcal{O}_K\Gamma$ mod p $\mathcal{O}_K\Gamma$, which is nilpotent. Thus for some positive integer n,

$$\mathfrak{r}_K^n \subseteq p\ \mathcal{O}_K\Gamma.$$

In particular, the $\mathfrak{r}_K$-adic and p-adic topologies of $\mathcal{O}_K\Gamma$ are equivalent.

Let C_Γ be the set of conjugacy classes of Γ, let KC_Γ be the K vector space on C_Γ, and let $\phi : K\Gamma \to KC_\Gamma$ be the K linear map which maps $\gamma \in \Gamma$ to $C(\gamma)$, the conjugacy class containing γ. Next we define $\Psi: KC_\Gamma \to KC_\Gamma$ to be the f-semi-linear map which maps $C(\gamma)$ to $C(\gamma^p)$. The usual p-adic logarithm extends to define a map

$$\log: 1 + \mathfrak{r}_K \to K\Gamma$$

where $\log(1 - r) = -\sum_1^\infty \frac{r^n}{n}$, for $r \in \mathfrak{r}_K$. When Γ is non-abelian, log is not a group homomorphism. With this in mind, we adjust this logarithm and define

$$L : 1 + \mathfrak{r}_K \to KC_\Gamma, \tag{1.1}$$

by $L = (p - \Psi) \circ \phi \circ \log$.

<u>Theorem 1.2</u>

(a) L is a homomorphism of groups

(b) $L(1 + \mathfrak{r}_K) \subseteq p\ \mathcal{O}_K C_\Gamma$ (that is to say L is an integral logarithm).

Part (a) is proved in this section and part (b) in section 2.

Let $\chi \in R_{\Gamma,p}$. Then χ defines a function $\chi: C_\Gamma \to Q_p^c$, by the rule $\chi(C(\gamma)) = \chi(\gamma)$. Therefore, by K-linearity, χ may be viewed as an element of $\mathrm{Hom}_K(KC_\Gamma, Q_p^c)$.

<u>Proposition 1.3</u>

$$\chi \circ L(1 - r) = \log(\mathrm{Det}(1 - r)(p\chi)) - \log(\mathrm{Det}(1 - r^f)(\psi_p\chi))$$

for $r \in \mathcal{R}_K$, $\chi \in R_{\Gamma,p}$ (Recall ψ_p is the pth Adams operation).

<u>Proof</u> We have the equality in $\mathrm{Hom}_K(KC_\Gamma, Q_p^c)$

$$\chi \circ (p - \Psi) = p\chi - f \circ \psi_p \chi$$

and so, in fact, it will suffice to show

$$\chi \circ \phi \circ \log(1 - r) = \log(\mathrm{Det}(1 - r)(\chi)) \tag{1.4}$$

for each character χ of Γ. Let $T: \Gamma \to GL_m(Q_p^c)$ be a representation of Γ with character χ. Conjugating T if necessary, we may assume $T(r)$ to be upper triangular, with diagonal entries $r_1, \ldots r_m$. By continuity we know that each $r_i \in \mathcal{P}^c$, and from the definition

$$\log(\mathrm{Det}(1 - r)(\chi)) = \log(\prod_i 1 - r_i) = - \sum_{n=1}^{\infty} \sum_{i=1}^{m} \frac{r_i^{\,n}}{n} .$$

$\sum_1^m r_i^{\,n}$ is the trace of $T(r^n) = \chi \circ \phi(r^n)$ and the result is shown.

<u>Corollary 1.5</u>

$\phi \circ \log$, and hence also L, is a homomorphism of Δ-modules.

<u>Proof</u> In order to show that $\phi \circ \log$ is a group homomorphism, we show

$$\phi \circ \log (xy) = \phi \circ \log(x) + \phi \circ \log(y)$$

for $x, y \in 1 + \mathcal{R}_K$. Since the characters of Γ span $\mathrm{Hom}_{Q_p^c}(Q_p^c C_\Gamma, Q_p^c)$, it suffices to show

$$\chi \circ \phi \circ \log(xy) = \chi \circ \phi \circ \log(x) + \chi \circ \phi \circ \log(y)$$

for $\chi \in R_{\Gamma,p}$; and this is immediate from (1.4). It is clear that $\phi \circ \log$ commutes with Δ action.

In the same way, observe that if $\mathrm{Det}(x) = 1$, for $x \in 1 + \mathcal{R}_K$, then $\phi \circ \log(x) = 0$; so that $\phi \circ \log$, and hence also L, factors through Det. In particular, there exists a unique Δ-homomorphism ν which makes the following triangle commute:

$$\begin{array}{ccc} 1 + \mathfrak{p}_K & \xrightarrow{\;L\;} & KC_\Gamma \\ {\scriptstyle \mathrm{Det}}\big\downarrow & \nearrow {\scriptstyle \nu} & \\ \mathrm{Det}(1 + \mathfrak{p}_K) & & \end{array}$$

Because the roots of unity of $\mathcal{O}_K^*$ with order prime to p are a full set of representatives for $\mathcal{O}_K\Gamma^*/1 + \mathfrak{p}_K$, we deduce that ν extends in a unique way to $\mathrm{Det}(\mathcal{O}_K\Gamma^*)$ with unchanged image. Henceforth, ν will be used to denote this extended homomorphism. Furthermore, when questions of change of group or basefield are involved, we write $\nu_{K,\Gamma}$ for ν, for clarity.

More generally $\mathrm{Det}(\mathcal{O}_K\Gamma^*)$ has finite index in $\mathrm{Hom}_{\Omega_K}(R_{\Gamma,p}, U_p)$ (cf. 1 (3.12)) and so, because KC_Γ is uniquely divisible, ν extends in a unique way to a homomorphism

$$\nu_o : \mathrm{Hom}_{\Omega_K}(R_{\Gamma,p}, U_p) \to KC_\Gamma.$$

From Proposition 1.3 it is immediate that for $h \in \mathrm{Hom}_{\Omega_K}(R_{\Gamma,p}, U_p)$

$$\chi(\nu_o(h)) = \log(h(p\chi).h(-\psi_p\chi)^f). \tag{1.6}$$

Let T denote the torsion subgroup of $\mathrm{Hom}_{\Omega_K}(R_{\Gamma,p}, U_p)$. It is obvious that $T \subseteq \mathrm{Ker}(\nu_o)$ since KC_Γ is torsion free. We now show:

Proposition 1.7

$$\mathrm{Ker}(\nu_o) = T.$$

Proof We argue by induction on $|\Gamma|$. To begin the induction, consider the case $\Gamma = \{1\}$ and suppose $g \in \mathrm{Ker}(\nu_o)$. Then by (1.6)

$$0 = \varepsilon_\Gamma(\nu_o(g)) = \log(g(\varepsilon_\Gamma)^p\, g(\varepsilon_\Gamma)^{-f}).$$

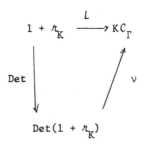

This implies $\log(g(\varepsilon_\Gamma)) = 0$ and so, because the usual p-adic logarithm on units of a local field has the roots of unity as its kernel, it follows that g is torsion.

We now need an elementary result from character theory.

<u>Lemma 1.8</u>

Let Λ be a central subgroup of Γ of order p and let $\chi \in R_{\Gamma,p}$. Then $\psi_p\chi$ is inflated from $\Gamma/_\Lambda$, i.e., may be lifted from $R_{\Gamma/\Lambda,p}$ by composition with the quotient map $\Gamma \to \Gamma/\Lambda$.

<u>Proof</u> Let α be a non-trivial abelian character of Λ. We wish to show that if an irreducible character occurs in $\psi_p\chi$, then it occurs in $\mathrm{Ind}_\Lambda^\Gamma\ \varepsilon_\Lambda$. Suppose θ is an irreducible character of Γ which occurs in $\mathrm{Ind}_\Lambda^\Gamma\ \alpha$. We show $(\theta, \psi_p\chi) = 0$.

Because θ occurs in $\mathrm{Ind}_\Lambda^\Gamma\alpha$, $\theta|_\Lambda = \theta(1)\alpha$ by Frobenius reciprocity. Let λ be a generator of Λ and choose a set of left coset representatives $\{\delta\}$ of Γ/Λ. Then, explicitly

$$\begin{aligned}(\theta,\psi_p\chi) &= \frac{1}{|\Gamma|}\sum_\delta \sum_{i=1}^{p} \theta(\delta\lambda^i)\ \psi_p\chi\ (\lambda^{-i}\ \delta^{-1}) \\ &= \frac{1}{|\Gamma|}\sum \chi(\delta^{-p})\ (\sum_1^p \theta(\delta\ \lambda^i)) = 0\end{aligned}$$

since $\theta(\delta\lambda) = \alpha(\lambda)\ \theta(\delta)$. □

We now complete the proof of Proposition 1.7. By the above it suffices to show that for $g \in \mathrm{Ker}(\nu_0)$ and $\chi \in R_{\Gamma,p}$, $\log(g(\chi)) = 0$. However, from (1.6) we know

$$0 = \chi(\nu(g)) = \log(g(p\chi)\ g(-\ \psi_p\chi)^f)$$

Using the lemma and the induction hypothesis applied to $\Gamma/_\Lambda$, we know $\log(g(\psi_p\chi)) = 0$, and the result is shown.

<u>Corollary 1.9</u>

$$\mathrm{Ker}(\nu) = \mathrm{Det}(\mathcal{O}_K\Gamma^*) \cap T.$$

Let $\mathfrak{a}_K = \mathrm{Ker}(\mathcal{O}_K\Gamma \to \mathcal{O}_K\Gamma^{ab})$. We shall now show

Proposition 1.10

$$\nu\Big|_{\mathrm{Det}(1 + \mathcal{a}_K)} \text{ is injective.}$$

Proof Suppose $g \in \mathrm{Ker}(\nu) \cap \mathrm{Det}(1 + \mathcal{a}_K)$. Since $\mathrm{Det}(1 + \mathcal{a}_K)$ is a pro-p-group g must have p-power order and so by 5 (1.1) $g = \mathrm{Det}(\mu\gamma)$ for some $\gamma \in \Gamma$, μ a root of unity of K^*. Since $g \in \mathrm{Det}(1 + \mathcal{a}_K)$, g must be 1 on all abelian characters of Γ. In particular $1 = g(\varepsilon_\Gamma) = \mu$; and so $1 = \mathrm{Det}(\gamma)(\chi) = \chi(\gamma)$ for all abelian characters of Γ. Thus $\gamma \in [\Gamma,\Gamma]$ and hence $\mathrm{Det}(\gamma) = 1$.

We now turn our attention to the image of ν. The main result needed is:

Proposition 1.11

$$\nu(\mathrm{Det}(1 + \mathcal{a}_K)) = p\phi(\mathcal{a}_K) .$$

We choose a central element $c \in \Gamma$ of order p, with $c = \alpha\beta\alpha^{-1}\beta^{-1}$ for some $\alpha, \beta \in \Gamma$. The existence of such an element c is assured by the lower-central series of Γ. As a first step towards proving the proposition we show:

Proposition 1.12

$$L(1 + (1 - c)\mathcal{O}_K\Gamma) = p\phi((1 - c)\mathcal{O}_K\Gamma)$$

Proof Clearly $\Psi \circ \phi((1 - c)\ \mathcal{O}_K\Gamma) = 0$, since $c^p = 1$. By successive approximation, it will suffice to show firstly that for $x \in \mathcal{O}_K\Gamma$

$$\phi \circ \log(1 + (1 - c)x) = \phi((1 - c)(x + y)) \tag{1.13}$$

with $y \in \mathcal{O}_K\Gamma$, and furthermore that if $x \in \mathcal{h}_K$ then we may take $y \in x\mathcal{h}_K$. Secondly we show

$$\phi((1 - c)\mathcal{O}_K\Gamma) = \phi((1 - c)\mathcal{h}_K). \tag{1.14}$$

Because $(1 - c)\mathcal{O}_K\Gamma/_{(1 - c)\mathcal{h}_K}$ is generated over $\mathcal{O}_K$ by $(1 - c)\beta$, we see that to prove (1.14) it suffices to observe

$$\phi((1 - c)\beta) = \phi(\beta - \alpha\beta\alpha^{-1}) = 0$$

In order to show (1.13) we need

Lemma 1.15

$$(1 - c)^p = p(1 - c)\lambda$$

with $\lambda \in \mathbb{Z}\langle c\rangle$ and furthermore $\lambda + 1$ lies in the radical of $\mathbb{Z}_p\langle c\rangle$.

Proof Clearly the left hand side is divisible by p by the Binomial theorem. Moreover it also has zero-augmentation, and so λ may indeed be chosen in $\mathbb{Z}\langle c\rangle$. Let χ be a non-trivial abelian character of $\langle c\rangle$. Since the image of λ modulo the radical of $\mathbb{Z}_p\langle c\rangle$ is determined by $\chi(\lambda) \bmod (1 - \zeta_p)$, using $\prod_{i=1}^{p-1} (1 - \zeta_p^i) = p$, it suffices to show

$$\frac{(1 - \zeta_p)^{p-1}}{\prod_1^{p-1} (1 - \zeta_p^i)} \equiv - 1 \bmod (1 - \zeta_p)$$

and this follows from the congruence

$$\frac{1 - \zeta_p^i}{1 - \zeta_p} \equiv i \bmod (1 - \zeta_p)$$

together with Wilson's theorem.

We now prove (1.13). Because all terms but first and p^{th} drop out mod $x(1 - c)\hbar_K$, it follows that

$$\log(1-(1 - c)x) = - \sum_{n=1}^{\infty} \frac{(1 - c)^n x^n}{n} \tag{1.16}$$

$$\equiv - (1 - c)(x + x^p\lambda) \bmod x(1 - c)\hbar_K.$$

(1.13) now follows. □

Corollary 1.17

For the moment we relax the condition that c be a commutator (but still insist c be central of order p). Because $x^p \equiv x^f \bmod \mathfrak{p}_K$, for $x \in \mathcal{O}_K\Gamma$, it follows from (1.13) and (1.16) by successive approximation, that

$$L\,(1 + (1 - c)\mathcal{O}_K\Gamma) \supseteq p(1 - c)\mathcal{O}_K(1 - f) + \phi(p(1 - c)\mathfrak{p}_K)\ .$$

In particular because $\mathcal{O}_K$ is a free Δ module $(\mathcal{O}_K\colon \mathcal{O}_K(1 - f) + p\mathcal{O}_K) = p$ and so

$$(\phi((1 - c)\mathcal{O}_K\Gamma)\ :\ L(1 + (1 - c)\mathcal{O}_K\Gamma)) \leqslant p.$$

We now prove Proposition 1.11 by induction on $|[\Gamma,\Gamma]|$. The result is trivial when $[\Gamma,\Gamma] = 1$. This begins our induction argument. We write $\tilde{\Gamma}$ for the quotient group $\Gamma/_{<c>}$. More generally, for the purposes of this proof, given an object x associated to Γ, we write the corresponding object for $\tilde{\Gamma}$ as $\tilde{x}$. The map $C_\Gamma \to C_{\tilde{\Gamma}}$ induces a K-linear homomorphism $\eta\colon KC_\Gamma \to KC_{\tilde{\Gamma}}$.

Lemma 1.18

$$\mathrm{Ker}(\eta) \cap \mathcal{O}_K C_\Gamma = \phi((1 - c)\mathcal{O}_K C_\Gamma)$$

Proof We write K for the left hand term in the statement of the lemma, and let K_1 (resp. K_2) be $\mathrm{Ker}(\mathcal{O}_K\Gamma \to \mathcal{O}_K C_\Gamma)$ (resp. $\mathrm{Ker}(\mathcal{O}_K\tilde{\Gamma} \to \mathcal{O}_K C_{\tilde{\Gamma}})$). Explicitly

$$K_1 = \sum_{\gamma,\delta\in\Gamma} \mathcal{O}_K(\gamma - \delta^{-1}\gamma\delta),$$

$$K_2 = \sum_{\gamma,\delta\in\tilde{\Gamma}} \mathcal{O}_K(\gamma - \delta^{-1}\gamma\delta).$$

Hence $\Gamma \to \tilde{\Gamma}$ induces a surjection $K_1 \to K_2$. The homomorphisms η and ϕ induce a commutative diagram with the two rows and the two right hand columns exact.

$$
\begin{array}{ccccccccc}
 & & & & 0 & & 0 & & \\
 & & & & \downarrow & & \downarrow & & \\
 & & & & K_1 & \longrightarrow & K_2 & \longrightarrow & 0 \\
 & & & & \downarrow & & \downarrow & & \\
0 \to & (1-c)\mathcal{O}_K\Gamma & & \longrightarrow & \mathcal{O}_K\Gamma & \longrightarrow & \mathcal{O}_K\tilde{\Gamma} & \longrightarrow & 0 \\
 & \downarrow{\scriptstyle \phi_*} & & & \downarrow{\scriptstyle \phi} & & \downarrow{\scriptstyle \tilde{\phi}} & & \\
0 \longrightarrow & K & & \longrightarrow & \mathcal{O}_K C_\Gamma & \xrightarrow{\ \eta\ } & \mathcal{O}_K C_{\tilde{\Gamma}} & \longrightarrow & 0 \\
 & & & & \downarrow & & \downarrow & & \\
 & & & & 0 & & 0 & &
\end{array}
$$

Thus ϕ_* is surjective, as we require. □

In order to complete our proof of Proposition 1.11 we show

$$L(1 + \mathcal{a}_K) \subseteq p\phi(\mathcal{a}_K), \tag{1.19}$$

$$L(1 + \mathcal{a}_K) \supseteq p\phi(\mathcal{a}_K). \tag{1.20}$$

Now by the induction hypothesis applied to $\tilde{\Gamma}$,

$$\tilde{L}(1 + \tilde{\mathcal{a}}_K) = p\tilde{\phi}(\tilde{\mathcal{a}}_K) = p\ \eta \circ \phi(\mathcal{a}_K). \tag{1.21}$$

From the definition of L (cf. (1.1)) it is immediate that for $a \in \mathcal{a}_K$ with image $\tilde{a} \in \tilde{\mathcal{a}}_K$, $\tilde{L}(1 + \tilde{a}) = \eta \circ L(1 + a)$.

<u>Proof of (1.19)</u> Let $a \in \mathcal{a}_K$. By (1.21) we can find $b \in \mathcal{a}_K$ so that $\tilde{L}(1 + \tilde{a}) = p\tilde{\phi}(\tilde{b})$, and so

$$L(1 + a) - p\phi(b) \in \mathrm{Ker}(\eta)$$

By Theorem 1.2(b), $L(1 + a) \in p\mathcal{O}_K C_\Gamma$, and so, by Lemma 1.18

$$L(1 + a) - p\phi(b) \in \mathrm{Ker}(\eta) \cap p\mathcal{O}_K C = p\phi((1 - c)\mathcal{O}_K\Gamma)) \subseteq p\phi(\mathcal{a}_K).$$

<u>Proof of 1.20</u> Let $a \in \mathcal{a}_K$. By (1.21) we can find some $b \in \mathcal{a}_K$ such that $\tilde{L}(1 + \tilde{b}) = p\tilde{\phi}(\tilde{a})$, and so

$$L(1 + b) - p\phi(a) \in \mathrm{Ker}(\eta).$$

Again by Lemma 1.18 and Theorem 1.2(b)

$$L(1 + b) - p\phi(a) \in p\phi((1 - c)\mathcal{O}_K\Gamma)$$

and the result then follows from proposition 1.12. □

Proposition 1.10 and 1.11 together imply

Theorem 1.22

The restriction of ν yields an isomorphism of Δ-modules

$$\mathrm{Det}(1 + \mathfrak{a}_K) \cong p\phi(\mathfrak{a}_K).$$

The homomorphisms $\Gamma \to \Gamma^{ab}$ and Det give rise to a commutative diagram:

$$\begin{array}{ccccccccc} 1 & \longrightarrow & 1 + \mathfrak{a}_K & \longrightarrow & \mathcal{O}_K\Gamma^* & \longrightarrow & \mathcal{O}_K\Gamma^{ab*} & \longrightarrow & 1 \\ & & & & \downarrow & & \downarrow & & \\ & & & & \mathrm{Det}(\mathcal{O}_K\Gamma^*) & \to & \mathrm{Det}(\mathcal{O}_K\Gamma^{ab*}) & \to & 1 \end{array}$$

The top row is exact, so that by 5 (1.2) and the Snake lemma we may deduce that

$$\mathrm{Det}(1 + \mathfrak{a}_K) = \mathrm{Ker}(\mathrm{Det}(\mathcal{O}_K\Gamma^*) \to \mathrm{Det}(\mathcal{O}_K\Gamma^{ab*})) \ .$$

From 5 (1.2) and Theorem 1.22 we derive the exact sequence of Δ-modules:

$$0 \to p\phi(\mathfrak{a}_K) \xrightarrow{\nu^{-1}} \mathrm{Det}(\mathcal{O}_K\Gamma^*) \longrightarrow \mathcal{O}_K\Gamma^{ab^*} \longrightarrow 1 \qquad (1.23)$$

This exact sequence is absolutely fundamental to the study of classgroups of group rings.

We next describe a trivial generalisation of (1.23) to Galois orders. This will be necessary, for technical reasons, in section 3. We fix an embedding of Δ into a finite group Λ. The ring $\mathcal{B} = \mathrm{Map}_\Delta(\Lambda, \mathcal{O}_K)$ is a Galois order with Galois group Λ. Since $\mathcal{B}$ is just a direct product of copies of $\mathcal{O}_K$ transitively permuted by Λ, we may immediately deduce from (1.23) the exact sequence of Λ-modules:

$$0 \to \phi(a_K) \otimes_{O_K} B \to \mathrm{Det}(B\Gamma^*) \to B\Gamma^{ab*} \to 1. \qquad (1.24)$$

Next we shall describe the cokernel of ν_K in $pO_K C_\Gamma$. The result we give is due to Robert Oliver cf. [O1].

Let I_Γ be the augmentation ideal of $\mathbb{Z}_p\Gamma$ and let $\omega_{K,\Gamma}$ be the composite of

$$O_K\Gamma^{ab} \underset{\mathrm{Tr}_{K/\mathbb{Q}_p}}{\to} \mathbb{Z}_p\Gamma^{ab} \to \frac{\mathbb{Z}_p + I_{\Gamma^{ab}}}{\mathbb{Z}_p + I^2_{\Gamma^{ab}}} \cong \Gamma^{ab}.$$

Theorem 1.25

The natural map $O_K C_\Gamma \to O_K\Gamma^{ab}$ composed with $\omega_{K,\Gamma}$ induces an isomorphism

$$\frac{O_K C_\Gamma}{\frac{1}{p}\,\mathrm{Im}(\nu_{K,\Gamma})} \cong \Gamma^{ab}.$$

Proof From Theorem 1.22 we have the isomorphism

$$\frac{O_K C_\Gamma}{\frac{1}{p}\,\mathrm{Im}(\nu_{K,\Gamma})} \cong \frac{O_K C_{\Gamma^{ab}}}{\frac{1}{p}\,\mathrm{Im}(\nu_{K,\Gamma^{ab}})}$$

and so, without loss of generality, we suppose Γ to be abelian. This being the case we may identify C_Γ and Γ, so that ϕ = id and $\nu = L$. The proof goes in two steps.

(A) Show $\left|O_K\Gamma/\frac{1}{p}\,\mathrm{Im}(\nu_{K,\Gamma})\right| \leq |\Gamma|$.

(B) Show $\frac{1}{p}\,\mathrm{Im}(\nu_{\mathbb{Q}_p,\Gamma}) \subseteq \mathbb{Z}_p + I^2_\Gamma$.

For then, because $\nu_{K,\Gamma}$ commutes with Δ-action, we know

$$\mathrm{Tr}_{K/\mathbb{Q}_p}(\mathrm{Im}(\nu_{K,\Gamma})) = \mathrm{Im}(\nu_{K,\Gamma} \circ N_{K/\mathbb{Q}_p})$$

$$\subseteq \mathrm{Im}(\nu_{\mathbb{Q}_p,\Gamma}),$$

and so $\omega_{K,\Gamma}(\mathrm{Det}(\mathcal{O}_K\Gamma^*)) = 1$.

First of all we prove (A) by induction on the group order.

Let $c \in \Gamma$ have order p. As previous we set $\tilde{\Gamma} = \Gamma/_{<c>}$ and, for an object x associated to Γ, we write $\tilde{x}$ for the corresponding object associated to $\tilde{\Gamma}$. The homomorphisms $\Gamma \to \tilde{\Gamma}$ and L induce a commutative diagram of Δ-modules with exact rows:

$$\begin{array}{ccccccc}
1 \to 1 + (1 - c)\mathcal{O}_K\Gamma & \longrightarrow & \mathcal{O}_K\Gamma^* & \longrightarrow & \mathcal{O}_K\tilde{\Gamma}^* & \longrightarrow & 1 \\
\Big\downarrow L_1 & & \Big\downarrow L & & \Big\downarrow \tilde{L} & & \\
0 \to p(1 - c)\mathcal{O}_K\Gamma & \longrightarrow & p\mathcal{O}_K\Gamma & \longrightarrow & p\mathcal{O}_K\tilde{\Gamma} & \longrightarrow & 1
\end{array}$$

By induction hypothesis $|\mathrm{Coker}(\tilde{L})| \leq |\tilde{\Gamma}|$. The result is then immediate by Corollary 1.17.

In order to prove (B) we need the following lemma which gives a good foretaste of the proof of Theorem 1.2(b).

<u>Lemma 1.26</u>

$$L(1 + I_\Gamma{}^2) \subseteq pI_\Gamma^2$$

<u>Proof</u> Let $x \in I_\Gamma^2$. By definition

$$L(1 - x) = -\sum_{n=1}^{\infty} \frac{px^n}{n} + \sum_{n=1}^{\infty} \frac{\Psi(x^n)}{n} .$$

We split up the first sum according as n is prime to p or not, and so

$$L(1 - x) = -\sum_{(n,p)=1} \frac{px^n}{n} + \sum_{n=1}^{\infty} \frac{x^{np} - \Psi(x^n)}{n} \tag{1.27}$$

Using the congruence $a^p \equiv a = \Psi(a) \bmod p\mathbb{Z}_p$ for $a \in \mathbb{Z}_p$, we deduce from the Binomial theorem that for $z \in I_\Gamma^2$

$$z^p = \Psi(z) + p\theta$$

$\theta \in I_\Gamma^2$. Raising to successive powers of p, we see that for $e \geq 0$

$$z^{p^{e+1}} \in \Psi(z^{p^e}) + p^{e+1} I_\Gamma^2.$$

If $n = p^e m$ with $(p,m) = 1$, then we set $z = x^m$ and the lemma follows from (1.27). □

We now observe that

$$\mathbb{Z}_p \Gamma^* = \mathbb{Z}_p{}^* (1 + I_\Gamma)$$

and that $1 + I_\Gamma / 1 + I_\Gamma^2$ is represented by the elements of Γ, which are of course torsion. Hence

$$\begin{aligned} \nu_{\mathbb{Q}_p,\Gamma}(\mathbb{Z}_p \Gamma^*) &= \nu_{\mathbb{Q}_p,\Gamma}(\mathbb{Z}_p{}^*) + \nu_{\mathbb{Q}_p,\Gamma}(1 + I_\Gamma^2) \\ &\subseteq p\mathbb{Z}_p + p\, I_\Gamma^2 \end{aligned}$$

by the above lemma. This then establishes (B).

§2. PROOF OF THEOREM 1.2(b)

The proof we shall give is essentially that of [T1] (cf. also Proposition 3.1 in [T2]). We shall include a number of improvements in style taken from [F7].

Let $r = \Sigma\, r_\gamma\, \gamma \in \mathcal{R}_K$. From the definition of L, we know

$$L(1 - r) = \phi\left(- \sum_{n=1}^{\infty} \frac{pr^n}{n} + \sum_{n=1}^{\infty} \frac{\Psi(r^n)}{n} \right)$$

We split up the terms in the first sum according as whether n is prime to p or not, and we get

$$L(1 - r) = - \sum_{(n,p)=1} \frac{p}{n} \phi(r^n) + \sum_{n=1}^{\infty} \frac{1}{n} (\phi(r^{np}) - \Psi \circ \phi(r^n))$$

Clearly the first term lies in $p\, \mathcal{O}_K C_\Gamma$. If p^{e-1} is the exact power of p dividing n, then in order to show that

$$\frac{1}{n} [\phi(r^{np}) - \Psi \circ \phi(r^n)] \in np\, \mathcal{O}_K C_\Gamma$$

it will suffice to show

$$\phi(y^{p^e}) - \Psi \circ \phi(y^{p^{e-1}}) \in p^e\, O_K C_\Gamma \tag{2.1}$$

for any $y \in O_K\Gamma$.

In order to show this we first establish a number of results in $A = O_K[\,.\ .\ X_\gamma\ .\ .\,]$, the free associative algebra on non-commuting indeterminates $\{X_\gamma\}_{\gamma\in\Gamma}$.

For $0 \leq j \leq e$ we define M_j to be the set of monomials in the $\{X_\gamma\}$ of length p^j. We write $O_K(M_j)$ for the free O_K-module on the elements of M_j. Let Σ be the cyclic group of order p^e, and let σ generate Σ. Then Σ acts on the elements of M_j, for each j, by the rule:

$$(X_{\gamma(1)} \cdots X_{\gamma(p^j)})^\sigma = X_{\gamma(p^j)} X_{\gamma(1)} \cdots X_{\gamma(p^j-1)}\ .$$

Let M_j/Σ denote the set of Σ orbits of M_j and let $\phi^*\colon O_K(M_j) \to O_K(M_j/\Sigma)$ be the O_K-linear maps induced by mapping an element of M_j to its Σ-orbit.

Let $\Psi^*\colon O_K(M_{e-1}) \to O_K M_e$ be the f semi-linear map induced by mapping an element of M_{e-1} to its pth-power. (2.1) will follow from:

<u>Proposition 2.2</u>

For $u \in O_K M_0$

$$\phi^*(u^{p^e} - \Psi^*(u^{p^{e-1}})) \in p^e O_K(M_e/\Sigma).$$

Before proving the proposition we show that it implies (2.1).

Let $a\colon A \to O_K\Gamma$ be the algebra homomorphism induced by mapping each $X_\gamma \mapsto \gamma$ for $\gamma \in \Gamma$. Then we obtain a commutative diagram:

$$\begin{array}{ccccc} O_K(M_{e-1}) & \xrightarrow{\ \Psi^*\ } & O_K(M_e) & \xrightarrow{\ \phi^*\ } & O_K(M_e/\Sigma) \\ \Big\downarrow a & & & \searrow\, \phi \circ a & \Big\downarrow a' \\ O_K\Gamma & & \xrightarrow{\ \Psi \circ \phi\ } & & O_K C_\Gamma \end{array}$$

with a' the O_K-linear map induced by mapping an orbit in $M_{e/\Sigma}$ to the conjugacy class of a evaluated on any element in that orbit.

For $y = \Sigma y_\gamma \gamma \in \mathcal{O}_K\Gamma$ as in (2.1), we choose

$$u = \Sigma y_\gamma X_\gamma .$$

Then, by the diagram,

$$\phi(y^{p^e}) - \Psi \circ \phi(y^{p^{e-1}}) = \phi \circ a(u^{p^e}) - \Psi \circ \phi \circ a(u^{p^{e-1}})$$

$$= a' \circ \phi^*(u^{p^e} - \Psi^*(u^{p^{e-1}}))$$

and so indeed (2.1) follows from the Proposition.

Proof of Proposition 2.2 Let us suppose

$$u^{p^{e-1}} = \sum_{m \in M_{e-1}} u_m m \qquad (2.3)$$

with $u_m \in \mathcal{O}_K$. For each d, $0 \leq d \leq e$, we write Σ_d for the unique subgroup of Σ of order p^d. A monomial $m \in M_{e-1}$ is stabilised by Σ_d when m has period p^{e-d} in the X_γ. Since m has length p^{e-1}, this means that $m = n^{p^{d-1}}$, for some $m \in M_{e-d}$. Notice, then, that the coefficient u_m in (2.3) must also be a p^{d-1} power, $u_m = b_m^{p^{d-1}}$ for some $b_m \in \mathcal{O}_K$.

We now suppose $\Sigma_{d(m)}$ to be the stabiliser of $m \in M_{e-1}$. Then, because $u_{m^\sigma} = u_m$ for $m \in M_{e-1}$, we see that

$$\phi^* \Psi(u^{p^{e-1}}) = \Sigma \Psi(b_m)^{p^{d(m)-1}} . p^{e-d(m)} \phi^*(m^p) \qquad (2.4)$$

where we sum over a set of representatives of the orbits $M_{e-1/\Sigma}$. Raising the equation (2.3) to the pth power:

$$u^{p^e} = \Sigma u_m^p m^p + \Sigma' v_n . n \qquad (2.5)$$

where Σ' extends over the monomials in M_e with trivial Σ-stabiliser.

We recall that $u_m = b_m^{p^{d(m)-1}}$, for some $b_m \in \mathcal{O}_K$, and that $u_{m^\sigma} = u_m$. Hence, on applying ϕ^* to (2.5),

$$\phi^*(u^{p^e}) \equiv \Sigma b_m^{p^{d(m)}} p^{e-d(m)} \phi^*(m^p) \bmod p^e \mathcal{O}_K(M_{e/\Sigma}). \qquad (2.6)$$

From the congruence $b^p \equiv b^f = \Psi(b) \bmod p\mathcal{O}_K$, using the Binomial theorem we obtain the higher congruence

$$b^{p^d} \equiv b^{p^{d-1}f} = \Psi(b^{p^{d-1}}) \bmod p^d\mathcal{O}_K.$$

Comparing (2.5) and (2.6) we see

$$\phi^*(u^{p^e}) \equiv \phi^* \Psi(u^{p^{e-1}}) \bmod p^e\mathcal{O}(M_{e/\Sigma}). \qquad \square$$

§3. Q-p-ELEMENTARY GROUPS

A finite group Γ is said to be Q-p-elementary if it is a semi-direct product of a p-group P by a normal cyclic group Σ with $(P,|\Sigma|) = 1$. For such a Γ we write $\Gamma = \Sigma \rtimes P$. Let s denote the order of Σ and let σ generate Σ.

Throughout this section K is a finite non-ramified extension of Q_p, and $\Delta = \mathrm{Gal}(K/Q_p)$. Our aim here is to extend the results of the previous sections, in order to describe $\mathrm{Det}(\mathcal{O}_K\Gamma^*)$ as a Δ-module, in the case when Γ is Q-p-elementary.

For an integer m dividing s, we define

$$\mathcal{O}_K[m] = \frac{\mathcal{O}_K[x]}{\Phi_m(x)\ \mathcal{O}_K[x]} \tag{3.1}$$

where $\Phi_m(x)$ is the mth cyclotomic polynomial; explicitly then $\mathcal{O}_K[m] \cong \mathcal{O}_K \otimes_{\mathbb{Z}} \mathbb{Z}[\eta_m]$ where η_m is a primitive m^{th} root of unity. $\mathcal{O}_K[m]$ is a Galois order in the sense of §1, and is a Δ-module in the obvious way.

Because $(s,p) = 1$, the map $\sigma \mapsto \prod_{m|s} \eta_m$ induces an isomorphism of rings

$$\mathcal{O}_K\Sigma \cong \prod_m \mathcal{O}_K[m]. \tag{3.2}$$

Conjugation yields a homomorphism $P \to \mathrm{Aut}(\Sigma)$, and so composition with a surjection $\Sigma \to \langle\eta_m\rangle$, gives rise to a homomorphism

$$a_m : P \to \mathrm{Aut}\langle\eta_m\rangle.$$

We let $H_m = \mathrm{Ker}\,(a_m)$ and $A_m = P/H_m$. The isomorphism (3.2) induces an isomorphism of rings

$$\mathcal{O}_K\Gamma \cong \prod_m \mathcal{O}_K[m] \circ P \tag{3.3}$$

where $\mathcal{O}_K[m] \circ P$ is the twisted group ring (with trivial cocycle) with

$$\alpha\eta\ \pi = \pi\ \alpha\eta^{a_m(\pi)}$$

for $\alpha \in \mathcal{O}_K$, $\eta \in \langle\eta_m\rangle$, $\pi \in P$. We now deduce:

Proposition 3.4

The isomorphism (3.3) yields an isomorphism of Δ-modules

$$\mathrm{Det}(\mathcal{O}_K\Gamma^*) \tilde{=} \prod_m \mathrm{Det}(\mathcal{O}_K[m] \circ P^*).$$

We put $K[m] = Q_p.\mathcal{O}_K[m]$, and we let $R_P^{(m)}$ denote the Grothendieck group of $(K[m] \circ P) \otimes_K Q_p^c$ modules. Since H_m acts trivially on $K[m]$, the subring $K[m]H$ in $K[m] \circ P$ is merely the standard group algebra. The Grothendieck group of $K[m]H \otimes_K Q_p^c$ modules is denoted by $R_H^{(m)}$.

Tensoring the isomorphism (3.3) by Q_p^c we obtain an isomorphism

$$R_{\Gamma,p} \tilde{=} \bigoplus_{m|s} R_P^{(m)}. \tag{3.5}$$

Viewing $R_H^{(m)}$ as a subgroup of $R_{\Sigma \rtimes H_m}$ via the projection $K\Sigma \rtimes H_m \to K[m]H_m$ we see that $R_H^{(m)}$ is generated by those characters of the form $\xi.\theta$ where θ is an irreducible character of H, ξ is an abelian character of Σ of order m, and $\xi.\theta$ is viewed as a character of $\Sigma \rtimes H_m$ in the natural way. Now $\mathrm{Ind}_{\Sigma \rtimes H_m}^{\Gamma}(\xi.\theta)$ is irreducible by Mackey's criterion, and so, viewing $R_P^{(m)}$ as a subgroup of R_Γ, we see that $R_P^{(m)}$ is generated by characters of the above form.

Isomorphism (3.5) induces a commutative diagram

$$\begin{array}{ccc} \mathrm{Hom}_{\Omega_K}(R_{\Gamma,p}, U_p) & \tilde{\to} & \prod_{m|s} \mathrm{Hom}_{\Omega_K}(R_p^{(m)}, U_p) \\ \uparrow & & \uparrow \\ \mathrm{Det}(\mathcal{O}_K\Gamma^*) & \tilde{\to} & \prod_{m|s} \mathrm{Det}(\mathcal{O}_K[m] \circ P^*). \end{array} \tag{3.6}$$

For the time being we fix a choice of $m|s$, and so frequently we shall write H,A in place of H_m, A_m. As in chapter 1, §4 restriction from P to H gives a homomorphism

$$\mathrm{Res}_P^H : \mathrm{Det}(\mathcal{O}_K[m] \circ P^*) \to \mathrm{Det}(\mathcal{O}_K[m]\ H^*).$$

Explicitly we choose a transversal $\{\rho_i\}$ of P/H and for $x \in \mathcal{O}_K[m] \circ P^*$, we set

$$x\rho_i = \Sigma\ \rho_j\ x_{i,j}$$

with $x_{i,j} \in \mathcal{O}_K[m]H$. Then $\mathrm{Res}_P^H(\mathrm{Det}(x)) = \mathrm{Det}(x_{i,j})$. Again as in 1 (4.11), for $\chi \in R_H^{(m)}$

$$\mathrm{Res}_P^H(\mathrm{Det}(x))(\chi) = \mathrm{Det}(x)\ (\mathrm{Ind}_{\Sigma \rtimes H}^{\Gamma}(\chi)).$$

Likewise composition with $\mathrm{Ind}_{\Sigma \rtimes H}^{\Gamma}$ induces a homomorphism μ

$$\mu\colon \mathrm{Hom}_{\Omega_K}(R_P^{(m)}, U_p) \to \mathrm{Hom}_{\Omega_K}(R_H^{(m)}, U_p).$$

Moreover the right hand group is an A-module by the rule $f^{\alpha}(\chi) = f(\chi^{\alpha})$ for $\alpha \in A$, $\chi \in R_H^{(m)}$, where χ^{α} denotes χ composed with conjugation by a representative in P of α. Explicitly, note that under this action

$$\mathrm{Det}(x)^{\alpha} = \mathrm{Det}(x^{\alpha}) \tag{3.7}$$

for $x \in \mathcal{O}_K[m]H^*$, where A acts on $\mathcal{O}_K[m]H$ by the homomorphism a_m on $\langle\eta_m\rangle$, by conjugation on H, but leaving $\mathcal{O}_K$ fixed.

To sum up, we have shown that restriction from P to H induces a commutative square

$$\begin{array}{ccc}
\mathrm{Hom}_{\Omega_K}(R_P^{(m)}, U_p) & \xrightarrow{\mu} & \mathrm{Hom}_{\Omega_K}(R_H^{(m)}, U_p)^A \\
\uparrow & & \uparrow \\
\mathrm{Det}(\mathcal{O}_K[m] \circ P^*) & \xrightarrow{\mathrm{Res}_P^H} & \mathrm{Det}(\mathcal{O}_K[m]H^*)^A .
\end{array} \tag{3.8}$$

Prior to (3.6) it was pointed out that

$$\mathrm{Ind}_{\Sigma \rtimes H}^{\Gamma} : R_H^{(m)} \to R_P^{(m)}$$

is surjective, whence μ and Res_P^H are injective. Our aim is to show Res_P^H is surjective so that we will then have shown:

Theorem 3.9

The isomorphism in (3.6), together with $\prod_m \mathrm{Res}_P^{H_m}$ gives rise to an isomorphism of Δ-modules

$$\mathrm{Det}(\mathcal{O}_K\Gamma^*) \cong \prod_{m|s} \mathrm{Det}(\mathcal{O}_K[m]\, H_m^*)^{A_m}.$$

Using the exact sequence (1.24) we will be able to obtain an exact sequence to describe the right hand terms (cf. (3.11)) and we shall then have our desired description of $\mathrm{Det}(\mathcal{O}_K\Gamma^*)$.

The norm homomorphism

$$N_A : \mathrm{Hom}_{\Omega_K}(R_H^{(m)}, U_p) \to \mathrm{Hom}_{\Omega_K}(R_H^{(m)}, U_p)^A$$

is defined by $N_A f = \prod_{\alpha \in A} f^{\alpha}$. It is immediate from Mackey's restriction formula that for $x \in \mathcal{O}_K[m]H^*$

$$\mathrm{Res}_P^H(\mathrm{Det}(x)) = \prod_{\alpha \in A} \mathrm{Det}(x)^{\alpha} = N_A(\mathrm{Det}(x)).$$

So we have inclusions

$$N_A(\mathrm{Det}(\mathcal{O}_K[m]H^*)) \subseteq \mathrm{Im}(\mathrm{Res}_P^H) \subseteq \mathrm{Det}(\mathcal{O}_K[m]H^*)^A$$

Hence, in order to prove Theorem 3.9, it suffices to show:

Proposition 3.10

$$N_A(\mathrm{Det}(\mathcal{O}_K[m]H^*)) = \mathrm{Det}(\mathcal{O}_K[m]H^*)^A.$$

Proof From the exact sequence (1.24) we have a commutative diagram:

$$\begin{array}{ccccccc} 0 \to p\phi(\mathfrak{a}_K) \otimes_{\mathbb{Z}_p} \mathbb{Z}_p[m] & \to & \mathrm{Det}(\mathcal{O}_K[m]H^*) & \to & \mathcal{O}_K[m]H^{ab^*} & \to & 1 \\ \downarrow \mathrm{Tr}_A & & \downarrow N_A & & \downarrow N'_A & & \\ 0 \to (p\phi(\mathfrak{a}_K) \otimes_{\mathbb{Z}_p} \mathbb{Z}_p[m])^A & \to & \mathrm{Det}(\mathcal{O}_K[m]H^*)^A & \to & (\mathcal{O}_K[m]H^{ab^*})^A & \to & 1 \end{array} \qquad (3.11)$$

The top row is exact by (1.24), and the bottom row is therefore exact since $\mathbb{Z}_p[m]$, whence also $p\phi(\mathfrak{a}_K) \otimes \mathbb{Z}_p[m]$, is $\mathbb{Z}_p A$ free. Here

(1) N'_A is the usual commutative algebra norm. Using powers of the Jacobson radical $\mathfrak{r}$ of $\mathbb{Z}_\ell H^{ab}$, we construct a filtration of $\mathbb{Z}_\ell H^{ab}$ ideals $\{h_i\}_1^\infty$ such that $h_1 = \mathfrak{r}$, $(h_i : h_{i+1}) = \ell$, $\mathfrak{r}h_i \subseteq h_{i+1}$. The surjectivity of N'_A then follows by successive approximation using the commutative diagram for $i \geq 1$

$$\begin{array}{ccc} \dfrac{1 + h_i\,\mathcal{O}_K[m]}{1 + h_{i+1}\,\mathcal{O}_K[m]} & \tilde{\to} & \dfrac{h_i\,\mathcal{O}_K[m]}{h_{i+1}\,\mathcal{O}_K[m]} \\ \downarrow N_A & & \downarrow \mathrm{Tr}'_A \\ \dfrac{1 + h_i\,\mathcal{O}_K[m]^A}{1 + h_{i+1}\,\mathcal{O}_K[m]^A} & \tilde{\to} & \dfrac{h_i\,\mathcal{O}_K[m]^A}{h_{i+1}\,\mathcal{O}_K[m]^A} \end{array}$$

observing that the trace from $h_i\,\mathcal{O}_K[m]$ to $h_i\,\mathcal{O}_K[m]^A$ is surjective (the former being $\mathbb{Z}_p A$ free) and using the fact that N'_A maps onto $(\mathcal{O}_K[m]\,H^{ab^*})^A$ mod $1 + \mathfrak{r}$, since this group has order prime to p.

(2) Tr_A is the trace map and this is surjective because $p\phi(\mathfrak{a}_K) \otimes \mathbb{Z}_p[m]$ is $\mathbb{Z}_p A$ free.

We now deduce that N_A is surjective and the proposition, whence also the theorem, is shown. □

Now we let Λ be a subgroup of $\Delta = \mathrm{Gal}(K/Q_p)$. From the above results we can very easily show:

<u>Theorem 3.12</u>

Let Γ be a Q-p-elementary group, then

$$\mathrm{Det}(\mathcal{O}_K\Gamma^*)^{\Lambda} = \mathrm{Det}(\mathcal{O}_{K^{\Lambda}}\Gamma^*).$$

<u>Proof</u> By Theorem 3.9 it suffices to show

$$(\mathrm{Det}(\mathcal{O}_K[m]H_m)^{A_m})^{\Lambda} = \mathrm{Det}(\mathcal{O}_{K^{\Lambda}}[m]H_m)^{A_m}$$

for each $m|s$. We now fix such an m, and once again surpress the subscript m. The inclusion $\supseteq$ is clear. Indeed this inclusion together with the lower exact sequence in (3.11) yields a commutative diagram:

$$\begin{array}{ccccccccc}
0 \to & (p\phi(\mathcal{a}_{Q_p}) \otimes_{\mathbb{Z}_p} \mathcal{O}_K[m]^{\Lambda})^A & \to & \mathrm{Det}(\mathcal{O}_K[m]H^*)^{\Lambda A} & \to & (\mathcal{O}_K[m]\, H^{ab^*})^{\Lambda A} & \to & 1 \\
& \uparrow \beta & & \uparrow & & \uparrow \alpha & & \\
0 \to & (p\phi(\mathcal{a}_{Q_p}) \otimes_{\mathbb{Z}_p} \mathcal{O}_{K^{\Lambda}}[m])^A & \to & \mathrm{Det}(\mathcal{O}_{K^{\Lambda}}[m]H^*)^A & \to & (\mathcal{O}_{K^{\Lambda}}[m]\, H^{ab^*})^A & \to & 1
\end{array}$$

Here the top row is exact, because $\mathcal{O}_K[m]$ is $\Lambda \times A$ free. However, α and β are both surjective since $\mathcal{O}_K[m]^{\Lambda} = \mathcal{O}_{K^{\Lambda}}[m]$, and so the result is shown.

7. SWAN MODULES AND CLASSGROUPS OF EXCEPTIONAL GROUPS

In section 1 we shall use the group logarithm to prove 3 (2.5) for the non-exceptional p-groups and, at the same time, we prove the following result conjectured by C.T.C. Wall in [W2].

Theorem 1

Let Γ be a finite group, let p be a prime number and let $z \in \mathbb{Z}_p \Gamma^*$ have the property that $\mathrm{Det}(z)(\chi) = 1$ for every irreducible character χ of Γ different from the identity character ε_Γ. Then

$$\mathrm{Det}(z)(\varepsilon_\Gamma) \equiv 1 \mod |\Gamma| \mathbb{Z}_p$$

In section 2 we prove the Fröhlich-Keating-Wilson theorem for dihedral and quaternion 2-groups and also Endo's theorem for semi-dihedral groups. This will then complete our proof of 3 Theorem 2.5.

§1. NON EXCEPTIONAL p-GROUPS AND WALL'S CONJECTURE

Let Γ be a finite p-group and let ρ denote the regular character of Γ. We fix an embedding $F \hookrightarrow Q_p^c$, this determines a map $U_p(F) \to U_p$, and we let e denote the composite homomorphism

$$\mathrm{Hom}_{\Omega_Q}(R_\Gamma, U_p(F)) \to \mathrm{Hom}_{\Omega_{Q_p}}(R_{\Gamma,p}, U_p) \xrightarrow{\nu_0} Q_p C \xrightarrow{\rho} Q_p \tag{1.1}$$

From 6 (1.2.b) we know

$$e(\mathrm{Det}\ U_p(\mathbb{Z}\Gamma)) \subseteq \rho(p\ \mathbb{Z}_p\ C) = p^{n+1}\ \mathbb{Z}_p \tag{1.2}$$

where $|\Gamma| = p^n$. On the other hand, if $h \in \mathrm{Hom}_{\Omega_Q}(R_\Gamma, \mathcal{O}_F^*)$, then by 6 (1.6)

$$e(h) = \log (h(p\rho)h(-\psi_p\rho)^f) = 0$$

since $h(\psi_p\rho)$ and $h(\rho)$ are elements of $\mathcal{O}_F^*$ which are Ω_Q fixed: namely ±1.

By the isomorphism 1 (3.11), e therefore induces a homomorphism:

$$e' : D(\mathbb{Z}\Gamma) \to Q_p \mod p^{n+1}\mathbb{Z}_p .$$

We shall now evaluate $e'(1 + p, \Sigma_\Gamma)$. We know from 1 (3.14) that

$$e'(1 + p, \Sigma_\Gamma) = e(f_{1+p}) \bmod p^{n+1}\mathbb{Z}_p$$

$$= \log(f_{1+p}(p\rho - \psi_p\rho)) \bmod p^{n+1}\mathbb{Z}_p$$

where $f_{1+p}(\chi) = (1 + p)^{(\varepsilon_\Gamma, \chi)}$ for $\chi \in R_\Gamma$. Thus

$$e'(1 + p, \Sigma_\Gamma) = (\varepsilon_\Gamma, p\rho - \psi_p\rho) \log(1 + p) \bmod p^{n+1}\mathbb{Z}_p .$$

Now
$$(\varepsilon_\Gamma, p\rho) = p,$$

$$(\varepsilon_\Gamma, \psi_p\rho) = \sum_{\gamma^p=1} 1$$

and $\log(1 + p) = p\delta_p$ with $\delta_p \in \mathbb{Z}_p^*$ (resp. $\delta_2 \in 2\mathbb{Z}_2^*$) if $p \neq 2$ (resp. $p = 2$). Before proceeding further, we need the following result from group theory

<u>Theorem 1.3</u>

(a) (Kulakoff cf. [Z]). Let $p \neq 2$ and let Γ be a non-cyclic p-group, then

$$\sum_{\gamma^p=1} 1 \equiv 0 \mod (p^2).$$

(b) (Alperin-Feit-Thompson cf. [L]). Let Γ be a non-exceptional, non-cyclic 2-group; then

$$\sum_{\gamma^2=1} 1 \equiv 0 \mod (4).$$

It now follows immediately that if $p \neq 2$ (resp. $p = 2$) and Γ is non-cyclic (resp. non-cyclic and non-exceptional) then $e'(1 + p, \Sigma_\Gamma)$ has order $p^{-1}|\Gamma|$ (resp. $\frac{1}{4}|\Gamma|$). However, from 3 (1.4) and 4 (2.1) we know that $T(\mathbb{Z}\Gamma)$ is a quotient of the p-part of $(\mathbb{Z}/_{|\Gamma|\mathbb{Z}})^*/\langle\pm1\rangle$, and this has order p^{n-1} (resp. $\frac{1}{4}|\Gamma|$). Part (a) of 3 Theorem (2.5) is now proved.

Next we prove Theorem 1 above. First we reduce from the general case to the p-group case. So we suppose that Γ is a finite group and we let Δ be a p-Sylow sub-group of Γ. Let z be as in the statement of the theorem and put

$$\mathrm{Det}(z_o) = \mathrm{Res}_\Gamma^\Delta(\mathrm{Det}(z))$$

cf. 1 (4.9). Then for $\chi \in R_\Delta$

$$\mathrm{Det}(z_o)(\chi) = \mathrm{Det}(z)(\mathrm{Ind}_\Delta^\Gamma\, \chi)$$

However, if χ is an irreducible character of Δ different from ε_Δ, then by Frobenius reciprocity

$$0 = (\varepsilon_\Delta, \chi) = (\varepsilon_\Gamma, \mathrm{Ind}_\Delta^\Gamma\, \chi)$$

$$1 = (\varepsilon_\Delta, \varepsilon_\Delta) = (\varepsilon_\Gamma, \mathrm{Ind}_\Delta^\Gamma\, \varepsilon_\Delta)$$

and so $\mathrm{Det}(z_o)$ satisfies the conditions of the theorem with Γ replaced by Δ. This completes our reduction since $|\Gamma|\mathbb{Z}_p = |\Delta|\mathbb{Z}_p$.

Next we suppose Γ to be abelian. From our hypothesis it is a triviality that in this case $z = 1 + w \sum_{\gamma\in\Gamma} \gamma$ for $w \in \mathbb{Z}_p$, and so indeed $\mathrm{Det}(z)(\varepsilon_\Gamma) = 1 + w|\Gamma|$.

From the above it certainly suffices to prove the theorem for non-cyclic p-groups Γ. In the next step we shall also suppose that Γ is not exceptional.

From (1.2)

$$e(\mathrm{Det}(z)) = \log(\mathrm{Det}(z)(p\rho - \psi_p\rho)) \in p^{n+1}\, \mathbb{Z}_p .$$

Because $\mathrm{Det}(z_o)$ is 1 on all irreducible characters of Γ different from ε_Γ, we see that for $\chi \in R_\Gamma$

$$\mathrm{Det}(z)(\chi) = \mathrm{Det}(z)(\varepsilon_\Gamma)^{(\varepsilon_\Gamma,\chi)},$$

and hence we deduce that

$$(p\rho - \psi_p\rho,\ \varepsilon_\Gamma)\log(\mathrm{Det}(z_o)(\varepsilon_\Gamma)) \in p^{n+1}\,\mathbb{Z}_p\ .$$

However, by Theorem 1.3 $(p\rho - \psi_p\rho,\ \varepsilon_\Gamma) \in p\,\mathbb{Z}_p^*$, and so by the elementary theory of the p-adic logarithm from $\mathbb{Z}_p^* \to \mathbb{Z}_p^+$, we deduce that

$$\mathrm{Det}(z)(\varepsilon_\Gamma) = \zeta(1 + p^n v)$$

with $v \in \mathbb{Z}_p$ and ζ a root of unity in $\mathbb{Z}_p^*$. It will now suffice to show $\zeta = 1$. Let $v_o = 1 + v \sum_{\gamma\in\Gamma} \gamma$ and put $z' = z\, v_o^{-1}$. Then $\mathrm{Det}(z')$ is ζ on ε_Γ and 1 on all other irreducible characters.

Next we choose an abelian character χ of Γ different from ε_Γ. Because $\chi - \varepsilon_\Gamma \in \mathrm{Ker}\, d_{p,\Gamma}$, from 4 (1.4) we know $\mathrm{Det}(z')(\chi - \varepsilon_\Gamma)$ lies in a pro-p-group; therefore ζ must have p-power order. In this case, by 5 (1.1), it follows that

$$\mathrm{Det}(z') = \begin{cases} \mathrm{Det}(\pm\gamma) & \text{if} \quad p = 2 \\ \mathrm{Det}(\gamma) & \text{if} \quad p \neq 2, \end{cases}$$

for some $\gamma \in \Gamma$.

If $p = 2$, then, since Γ is non-cyclic, there exists an abelian character χ of Γ, $\chi \neq \varepsilon_\Gamma$, such that $\chi(\gamma) = 1$. Thus $\mathrm{Det}(-\gamma)(\chi) = -1$ which would contradict the hypothesis. So in all cases we have $\mathrm{Det}(z') = \mathrm{Det}(\gamma)$.

Again, by hypothesis, we know $\phi(\gamma) = \mathrm{Det}(\gamma)(\phi) = 1$ for all abelian characters ϕ of Γ different from ε_Γ. Thus $\gamma \in [\Gamma,\Gamma]$ and so $\mathrm{Det}(\gamma) = 1$, as required.

In order to complete the proof of Theorem 1 we shall suppose that $p = 2$ and that Γ is exceptional. The proof in this case will follow easily from Proposition 1.4 below. This proposition exhibits a rather extraordinary determinantal relationship for units in semidihedral group rings, and then uses the fact that dihedral and quaternion 2-groups embed into a semidihedral 2-group.

Proposition 1.4

Let $|\Gamma| = 2^n$, let E be the field obtained by adjoining the 2^{n-1} st roots of unity to Q_2. We denote by Tr the trace from E to Q_2.

(a) Let Γ be a semi-dihedral group of order 2^n with presentation as given in 3 (2.3). For an abelian character θ of $<\xi>$, we write θ^* for $\mathrm{Ind}_{<\xi>}^{\Gamma}\ \theta$. Let χ be a faithful abelian character of $<\xi>$, let Φ be the group of non-faithful abelian characters of $<\xi>$ and put $\varepsilon = \varepsilon_{<\xi>}$. Then for each $z \in \mathbb{Z}_2\Gamma^*$, we have the congruence mod $4\mathbb{Z}_2$

$$\begin{aligned} &2^{2-n}\ \mathrm{Tr}\ (\mathrm{Det}(z)(\chi^*)) - \mathrm{Det}(z)(\det(\chi^*)) \\ \equiv\ & \qquad\qquad\qquad\qquad\qquad\qquad\qquad\qquad (1.5) \\ &2^{2-n} \sum_{\phi\in\Phi} \mathrm{Det}(z)(\phi^*) - \mathrm{Det}(z)(\det(\varepsilon^*)) \end{aligned}$$

(b) Let Γ be a dihedral group of order 2^n with presentation as given in 3 (2.1). For an abelian character θ of $<\alpha>$, we write θ^* for $\mathrm{Ind}_{<\alpha>}^{\Gamma}\ \theta$. Let χ be a faithful abelian character of $<\alpha>$, let Φ be the group of non-faithful abelian characters of $<\alpha>$ and let $\varepsilon = \varepsilon_{<\alpha>}$. Then the congruence (1.5) holds for each $z \in \mathbb{Z}_2\Gamma^*$.

(c) Let Γ be a quaternion 2-group with presentation as given in 3 (2.2). For an abelian character θ of $<\sigma>$, we write θ^* for $\mathrm{Ind}_{<\sigma>}^{\Gamma}\theta$. Let χ be a faithful abelian character of $<\sigma>$ and let Φ be the group of non-faithful abelian characters of $<\sigma>$ and put $\varepsilon = \varepsilon_{<\sigma>}$. Then the congruence (1.5) holds for each $z \in \mathbb{Z}_2\Gamma^*$.

Before proving the proposition, we will show that it will establish Theorem 1 for exceptional 2-groups; thereby completing the proof of Theorem 1.

Suppose that Γ is semidihedral (resp. dihedral, resp. quaternion) with order 2^n, and that z is as given in the statement of Theorem 1 with $\mathrm{Det}(z)(\varepsilon_\Gamma) = r$. From (1.5) we can immediately read off the congruences:

$$0 \equiv (2^{n-2} - 1 + r) - 2^{n-2} \bmod 2^n\ \mathbb{Z}_2 \ \text{ if } \Gamma \text{ semi-dihedral or dihedral}$$

$$2^{n-2}(1 - r) \equiv r - 1 \bmod 2^n\ \mathbb{Z}_2 \ \text{ if } \Gamma \text{ quaternion.}$$

Hence in all cases $r \equiv 1 \bmod 2^n\mathbb{Z}_2$, as required.

<u>Proof of Proposition 1.4</u> The dihedral (resp. quaternion) group of order 2^{n-1} embeds into the semi-dihedral group of order 2^n by the homomorphism induced by

$$\alpha \mapsto \xi^2 \qquad \beta \mapsto \eta$$

$$\text{resp.} \qquad \sigma \mapsto \xi^2 \qquad \tau \mapsto \eta\xi$$

using the presentations 3 (2.1), (2.2), (2.3). Thus parts (b) and (c) follow from part (a) by restriction.

We now prove part (a). Let $z = \lambda + \mu\eta$ with $\lambda, \mu \in \mathbb{Z}_2 \langle\xi\rangle$. For any abelian character θ of $\langle\xi\rangle$ it is easily calculated that

$$\mathrm{Det}(z)(\theta^*) = \mathrm{Det}(\lambda)(\theta^*) - \mathrm{Det}(\mu)(\theta^*)$$

$$\mathrm{Det}(z)(\det(\theta^*)) = \det(\theta^*)(z) = \det(\theta^*)(\lambda) - \det(\theta^*)(\mu).$$

Therefore, by linearity, we are reduced to the case $z = \sum_{i=1}^{2^{n-1}} a_i \xi^i$, with $a_i \in \mathbb{Z}_2$. For such z, we will prove the following two equalities:

$$2^{2-n}\,\mathrm{Tr}\,(\mathrm{Det}(z)(\chi^*)) = \Sigma\, a_i^{\,2}\, (-1)^i + \sum_{\substack{i \equiv j \bmod 2^{n-2} \\ i>j}} (-1)^{i+1}\, 2a_i a_j \tag{1.6}$$

$$2^{2-n} \sum_{\phi\in\Phi} \mathrm{Det}(z)(\phi^*) = \Sigma\, a_i^{\,2} + \sum_{\substack{i \equiv j \bmod 2^{n-2} \\ i>j}} 2a_i a_j \tag{1.7}$$

Before proving these two results, we show that they prove Proposition 1.4 (a).

Since $\det(\chi^*)(\xi) = -1$ and $\det(\varepsilon^*)(\xi) = 1$, we see that

$$\mathrm{Det}(z)(\det(\chi^*)) = \Sigma\, a_i\, (-1)^i \tag{1.8}$$

$$\mathrm{Det}(z)(\det(\varepsilon^*)) = \Sigma\, a_i\ . \tag{1.9}$$

Congruence (1.5) then follows from the congruence mod $4\mathbb{Z}_2$

$$(a_i{}^2 - a_i)(-1)^i \equiv a_i{}^2 - a_i \qquad \pm 2\, a_i a_j \equiv 2a_i a_j \;.$$

From Mackey's restriction formula we know that for an abelian character θ of $<\xi>$

$$\begin{aligned}\mathrm{Det}(z)(\theta^*) &= \mathrm{Det}(z)(\mathrm{Res}_{\Gamma}^{<\xi>}\, \theta) \\ &= \theta(z.\eta z\eta) \\ &= \theta(\sum_{i,j} a_i a_j\, \xi^{i+j(-1+2^{n-2})}) \end{aligned} \qquad (1.10)$$

On the one hand we know

$$\mathrm{Tr}(\chi(\xi^i)) = \begin{cases} 2^{n-2}\, \chi(\xi^i) & \text{if} \quad 2^{n-2} | i \;, \\ 0 & \text{otherwise} \end{cases} \qquad (1.11)$$

and by orthogonality

$$\sum_{\phi \in \Phi} \phi(\xi^i) = \begin{cases} 2^{n-2} & \text{if} \quad 2^{n-2} | i \;, \\ 0 & \text{otherwise} \;. \end{cases} \qquad (1.12)$$

(1.11) (resp. (1.12)) together with (1.10) now give (1.6) (resp. (1.7)). [Note that for $0 \leq j < i < 2^{n-1}$, $i \equiv j \bmod 2^{n-2}$, if, and only if, $i = j + 2^{n-2}$].

§2. EXCEPTIONAL 2-GROUPS

In general we shall write Δ_n (resp. H_n, resp. Σ_n) for the dihedral group (resp. quaternion group, resp. semi-dihedral group of order 2^n for $n \geq 2$ (resp. $n \geq 3$, resp. $n \geq 4$). As a notational convenience Δ_1 will be used to denote the group of order 2.

In this section we prove 3 (2.5.b). The proof will follow from:

Theorem 2.1

(a) For $n \geq 2$ $\quad D(\mathbb{Z}\Delta_n) = \{1\}$.

(b) For $n \geq 3$ (resp. $n \geq 4$)

$$|D(\mathbb{Z} H_n)| \leq 2 \quad \text{resp.} \quad |D(\mathbb{Z} \Sigma_n)| \leq 2.$$

(c) $T(\mathbb{Z} H_3) \cong \langle \pm 1 \rangle$.

<u>Remark</u> (c) was first shown by J. Martinet.

We begin by showing that Theorem 2.1 does indeed imply 3 (2.5.b). In the dihedral case this is clear. On the other hand the quaternion 2-groups H_n and the semi-dihedral 2-groups Σ_n all contain a subgroup isomorphic to H_3. Thus by 3 (1.7) we know $T(\mathbb{Z} H_n)$ and $T(\mathbb{Z} \Sigma_n)$ map onto $T(\mathbb{Z} H_3)$. Therefore from parts (b) and (c) of the above theorem

$$D(\mathbb{Z} H_n) \cong T(\mathbb{Z} H_n) \cong \langle \pm 1 \rangle,$$

$$D(\mathbb{Z} \Sigma_n) \cong T(\mathbb{Z} \Sigma_n) \cong \langle \pm 1 \rangle.$$

Let η_n be a primitive 2^{n-1} st root of unity, and let $C_2 = \langle c \rangle$ be a cyclic group of order 2. We let $\mathbb{Z}[\eta_n]^+$ (resp. $\mathbb{Z}[\eta_n]^-$) be the ring of integers of $Q(\eta_n + \eta_n^{-1})$ (resp. $Q(\eta_n - \eta_n^{-1})$). It must be pointed out that, as always, the subscript 2 on a $\mathbb{Z}$-algebra means completion at 2. The twisted group rings

$$(\mathbb{Z}[\eta_n], C_2)^1, \quad (\mathbb{Z}[\eta_n], C_2)^2, \quad (\mathbb{Z}[\eta_n], C_2)^3$$

all have underlying module $\mathbb{Z}[\eta_n]C_2$, and have multiplication

$$c^2 = 1, \quad \eta_n c = c\eta_n^{-1} \quad \text{in} \quad (\mathbb{Z}[\eta_n], C_2)^1, \quad n \geq 1, \tag{2.2}$$

$$c^2 = -1, \quad \eta_n c = c\eta_n^{-1} \quad \text{in} \quad (\mathbb{Z}[\eta_n], C_2)^2, \quad n \geq 3,$$

$$c^2 = 1, \quad \eta_n c = -c\eta_n^{-1} \quad \text{in} \quad (\mathbb{Z}[\eta_n], C_2)^3, \quad n \geq 4.$$

Then $\mathbb{Z}\Delta_n$ (resp. $\mathbb{Z}H_n$, resp. $\mathbb{Z}\Sigma_n$) maps onto $(\mathbb{Z}[\eta_n], C_2)^1$ (resp. $(\mathbb{Z}[\eta_n], C_2)^2$, resp. $(\mathbb{Z}[\eta_n], C_2)^3$) in a natural way.

We shall frequently need the following result:

Lemma 2.3

Let p be the maximal ideal of $\mathbb{Z}_2[\eta_n]^+$ (resp. $\mathbb{Z}_2[\eta_n]^-$) and let N, Tr be the norm and trace from $\mathbb{Z}_2[\eta_n]$ to $\mathbb{Z}_2[\eta_n]^+$ (resp. $\mathbb{Z}_2[\eta_n]^-$). Then for $r \geq 1$

$$N(1 + p^r \mathbb{Z}_2[\eta_n]) \supseteq 1 + p^{r+1} .$$

Proof Since Tr maps the maximal ideal of $\mathbb{Z}_2[\eta_n]$ onto p, the result follows by successive approximation, using the trivial congruence:

$$N(1 + x) = 1 + Tr(x) \bmod N(x)\ \mathbb{Z}_2[\eta_n]$$

for $x \in (1 - \eta_n) p\, \mathbb{Z}_2[\eta_n]$.

Proposition 2.4

$$D((\mathbb{Z}[\eta_n], C_2)^i) = 1$$

for i = 1,2,3 and for all permitted n.

Remark All these results follow from Theorem 3.4 [Wi].

Proof We prove for the case i = 1 - the remaining cases being virtually identical. For brevity, let $\Lambda(n) = (\mathbb{Z}[\eta_n], C_2)^1$. When n = 1, 2 this is 4 (2.3); so we suppose $n \geq 3$. Then $\Lambda(n) \otimes_{\mathbb{Z}} Q^c$ has one irreducible character up to Galois conjugacy (corresponding to a faithful irreducible character of Δ_n). We denote such a module by χ. Using the isomorphism 1 (3.11) and evaluating on χ, we obtain an isomorphism

$$D(\Lambda(n)) \cong \frac{(\mathbb{Z}_2[\eta_n]^+)^*}{(\mathbb{Z}[\eta_n]^+)^*\ Det(\Lambda(n)_2^*)(\chi)} .$$

Therefore it will suffice to show

$$Det(\Lambda(n)_2^*)(\chi) = (\mathbb{Z}_2[\eta_n]^+)^* \qquad (2.5)$$

Now, for $z \in \mathbb{Z}_2[\eta_n]$, by Mackey restriction,

$$\mathrm{Det}(1 + (1 - \eta_n)z)(\chi) = N(1 + (1 - \eta_n)z)$$

up to Galois conjugacy. And by Lemma 2.3

$$N(1 + (1 - \eta_n)\mathbb{Z}_2[\eta_n]) \supseteq 1 + p^2.$$

So now we need only observe that, up to Galois conjugacy,

$$\mathrm{Det}(1 + (1 - \eta_n)c)(\chi) = 1 - (1 - \eta_n)(1 - \eta_n^{-1}).$$

(2.5) is now shown.

Before proving Theorem 2.1.a we need some results concerning the distribution of units at 2. For any (possibly infinite) prime q of $\mathbb{Z}[\eta_n]^+$, we let $(\, , \,)_q$ be the Hilbert symbol of that prime. In [We], Weber showed:

<u>Proposition 2.6</u>

Let $n \geq 3$ and let q_i $1 \leq i \leq 2^{n-2}$ denote the infinite primes of $Q(\eta_n + \eta_n^{-1})$. There exist units η_i for $1 \leq i \leq 2^{n-3}$ such that η_i is positive at all q_j for $j \neq i$ and negative at q_i.

<u>Corollary 2.7</u>

Let p be the prime of $Q(\eta_n + \eta_n^{-1})$ over 2. Then

$$(\eta_i, \eta_j)_p = (-1)^{\delta_{i,j}} .$$

<u>Proof</u> By the product formula

$$(\eta_i, \eta_j)_p = \prod_{\mathfrak{r}} (\eta_i, \eta_j)_{\mathfrak{r}}$$

where $\mathfrak{r}$ runs through all primes of $Q(\eta_n + \eta_n^{-1})$ away from 2. However, for a finite such prime $\mathfrak{r}$ $(\eta_i, \eta_j)_{\mathfrak{r}} = 1$. Thus the result follows from the proposition.

Reduction mod 4 yields a homomorphism

$$r: \frac{(\mathbb{Z}[\eta_n]^+)^*}{((\mathbb{Z}[\eta_n]^+)^*)^2} \to \frac{(\mathbb{Z}_2[\eta_n]^+)^*}{((\mathbb{Z}_2[\eta_n]^+)^*)^2\,(1 + 4\mathbb{Z}_2[\eta_n]^+)} \qquad (2.8)$$

Lemma 2.9 r is surjective.

Proof Both groups in (2.8) are $\mathbb{F}_2$ vector spaces of dimension 2^{n-3}. Thus it suffices to show r is injective. Assume $u \in (\mathbb{Z}[\eta_n]^+)^*$ maps to 1 under r. Then, by a result of Hecke cf. [He], $\mathbb{Q}(\eta_n + \eta_n^{-1}, \sqrt[2]{u}) / \mathbb{Q}(\eta_n + \eta_n^{-1})$ is non-ramified, so that $(u,v)_p = 1$ for all $v \in (\mathbb{Z}[\eta_n]^+)^*$. However, from (2.6) we know that $(\ , \ )_p$ defines a non-degenerate form on $(\mathbb{Z}[\eta_n]^+)^* / ((\mathbb{Z}[\eta_n]^+)^*)^2$, and therefore $u \in ((\mathbb{Z}[\eta_n]^+)^*)^2$.

By Nakayama's lemma we deduce

Corollary 2.10
$(\mathbb{Z}[\eta_n]^+)^*$ maps onto $(\mathbb{Z}_2[\eta_n]^+)^*$ mod 4.

Proof of (2.1.a) We show $D(\mathbb{Z}\Delta_n) = 1$ for $n \geq 1$, and we argue by induction on n. The induction is started by 4 (2.3). Let $\Delta_n \to \Delta_{n-1}$ be a natural quotient map. Then we have an inclusion

$$\mathbb{Z}\Delta_n \hookrightarrow \Lambda(n) \times \mathbb{Z}\Delta_{n-1}.$$

Proposition 2.4 and the induction hypothesis imply that $D(\Lambda(n) \times \mathbb{Z}\Delta_{n-1}) = \{1\}$. Thus, using the isomorphism of 1 (3.11)

$$D(\mathbb{Z}\Delta_n) = \operatorname{Ker}\left(D(\mathbb{Z}\Delta_n) \to D(\Lambda(n) \times \mathbb{Z}\Delta_{n-1})\right)$$

$$\cong \frac{\operatorname{Hom}_{\Omega_Q}(R_{\Delta_n}, \mathcal{O}_F^*) \operatorname{Det}(\Lambda(n)_2^* \times \mathbb{Z}_2\Delta_{n-1}^*)}{\operatorname{Hom}_{\Omega_Q}(R_{\Delta_n}, \mathcal{O}_F^*) \operatorname{Det}(\mathbb{Z}_2\Delta_n^*)} \qquad (2.11)$$

With the notation of 3 (2.1) we know that

$$(\alpha^{2^{n-2}} - 1)\mathbb{Z}\Delta_n = \operatorname{Ker}(\mathbb{Z}\Delta_n \to \mathbb{Z}\Delta_{n-1}). \qquad (2.12)$$

Let K denote the subgroup of those homomorphisms in the denominator of (2.11) which vanish under the map induced by $\Delta_n \to \Delta_{n-1}$ (i.e., which vanish on $\operatorname{Inf}_{\Delta_{n-1}}^{\Delta_n} R_{\Delta_{n-1}}$).

First, suppose n = 2. Then $\Lambda(2) = \mathbb{Z}C_2$. Because both numerator and denominator in (2.11) have common image under projection to Δ_1, on applying the projection $\pi: Q\Delta_2 \to QC_2$ we obtain the isomorphism:

$$D(\mathbb{Z}\Delta_2) \cong \frac{\mathrm{Hom}(R_{C_2}, \pm 1)\ \mathrm{Det}(\mathbb{Z}_2 C_2^*)}{\pi(K)}$$

However, by (2.12) $\pi(K) \supseteq \mathrm{Hom}(R_{C_2}, \pm 1)\ \mathrm{Det}(1 + 2\mathbb{Z}_2 C_2)$ and since $\mathbb{F}_2 C^*_2 = \langle c \rangle$, we see that

$$\mathrm{Hom}(R_{C_2}, \pm 1)\ \mathrm{Det}(\mathbb{Z}_2 C_2^*) = \mathrm{Hom}(R_{C_2}, \pm 1)\ \mathrm{Det}(1 + 2\mathbb{Z}_2 C_2).$$

This, then, proves the result when n = 2. Now suppose $n \geqslant 3$. Again, because numerator and denominator in (2.11) have common image under projection to Δ_{n-1}, evaluation on χ of the kernels of these two groups under projection gives an isomorphism

$$D(\mathbb{Z}\Delta_n) \cong \frac{(\mathbb{Z}_2[\eta_n]^+)^*}{K(\chi)} \tag{2.13}$$

Note that here we have used (2.5) to obtain the numerator in (2.13). However, from (2.12) we know

$$K(\chi) \supseteq (\mathbb{Z}[\eta_n]^+)^*\ \mathrm{Det}(1 + 2\Lambda(n)_2)(\chi). \tag{2.14}$$

While from Lemma (2.3)

$$\mathrm{Det}(1 + 2\Lambda(n)_2)(\chi) \supseteq \mathrm{Det}(1 + 2\mathbb{Z}_2[\eta_n])(\chi) = N(1 + 2\mathbb{Z}_2[\eta_n]) \supseteq 1 + 2p$$

Theorem 2.1.a now follows immediately from (2.13), (2.14) and (2.10).

We remark that exactly the same argument as that used above with Δ_n replaced by H_n (resp. Σ_n) shows that $D(\mathbb{Z}H_n)$ resp. $D(\mathbb{Z}\Sigma_n)$ is a quotient of

$$A = \frac{(\mathbb{Z}_2[\eta_n]^+)^*}{(\mathbb{Z}[\eta_n]^+)^*_+\ (1 + 2p)} \quad \text{resp.} \quad B = \frac{(\mathbb{Z}_2[\eta_n]^-)^*}{(\mathbb{Z}[\eta_n]^-)^*\ (1 + 2p)}$$

Here $(\mathbb{Z}[\eta_n]^+)^*_+$ denotes the subgroup of totally positive units in $\mathbb{Z}[\eta_n]^+$. This arises because the irreducible faithful characters of H_n are symplectic.

Proposition 2.15

$D(\mathbb{Z}H_n)$ and $D(\mathbb{Z}\Sigma_n)$ are annihilated by 2.

Proof Since any square in $(\mathbb{Z}[\eta_n]^+)^*$ is totally positive, A is seen to be killed by 2, by Corollary (2.10).

In the semi-dihedral case we observe that the residue classes of $(\mathbb{Z}_2[\eta_{n-1}]^+)^*$ mod (2) and $((\mathbb{Z}_2[\eta_n]^-)^*)^2$ mod (2) coincide in $(\mathbb{Z}_2[\eta_n]^-)^*$ mod (2). Essentially this is because

$$(\eta_n - \eta_n^{-1})^2 \equiv \eta_{n-1} + \eta_{n-1}^{-1} \bmod 2$$

while $\eta_n - \eta_n^{-1}$ resp. $\eta_{n-1} + \eta_{n-1}^{-1}$ is a local uniformising parameter in $\mathbb{Z}_2[\eta_n]^-$ resp. $\mathbb{Z}_2[\eta_{n-1}]^+$. In other words, given a class $c \in \mathbb{Z}_2[\eta_n]^-$ mod $2p$, by (2.10) we can find $u \in (\mathbb{Z}[\eta_{n-1}]^+)^*$ such that $c^2u \equiv \pm 1 \bmod 2p$. This shows that B is killed by 2.

Proof of (2.1.b) Let Γ be either H_n or Σ_n and let Δ_{n-1} be the natural dihedral quotient of order 2^{n-1} obtained by factoring out by $\sigma^{2^{n-2}}$ resp. $\xi^{2^{n-2}}$. Let Λ denote the twisted group ring $(\mathbb{Z}[\eta_n], C_2)^2$ resp. $(\mathbb{Z}[\eta_n], C_2)^3$ if Γ is H_n resp. Σ_n. In exactly the same way as for (2.5), we obtain the equality:

$$\mathrm{Det}(\Lambda_2^*)(\chi) = \begin{cases} \mathbb{Z}_2[\eta_n]^+ & \Gamma = H_n, \\ \mathbb{Z}_2[\eta_n]^- & \Gamma = \Sigma_n, \end{cases}$$

where χ denotes an irreducible $\Lambda \otimes \mathbb{Q}^c$ module.

The inclusion $i: \mathbb{Z}\Gamma \hookrightarrow \mathbb{Z}\Delta_{n-1} \times \Lambda$ induces a homomorphism

$$i_* : D(\mathbb{Z}\Gamma) \to D(\mathbb{Z}\Delta_{n-1} \times \Lambda).$$

However, by (2.1.a) and Proposition (2.4)

$$D(\mathbb{Z}\,\Delta_{n-1} \times \Lambda) = \{1\}$$

and therefore, using the isomorphism 1 (3.11), we have

$$D(\mathbb{Z}\,\Gamma) = \frac{\mathrm{Hom}^{+}_{\Omega_Q}(R_\Gamma, \mathcal{O}_F^*)\,\mathrm{Det}(\mathbb{Z}_2\,\Delta^*_{n-1} \times \Lambda^*_2)}{\mathrm{Hom}^{+}_{\Omega_Q}(R_\Gamma, \mathcal{O}_F^*)\,\mathrm{Det}(\mathbb{Z}_2\Gamma^*)} \tag{2.16}$$

Let K denote the kernel of the denominator in (2.16) under the map π induced by the projection $\mathbb{Z}\,\Gamma \to \Lambda$. (If χ is an irreducible $\Lambda \otimes_{\mathbb{Z}} Q^c$ module, then π corresponds to evaluation on χ). Since both numerator and denominator in (2.16) have the same image under π, we obtain an isomorphism

$$D(\mathbb{Z}\,\Gamma) \cong \frac{\mathrm{Hom}_{\Omega_Q}(R_{\Delta_{n-1}}, \mathcal{O}_F^*)\,\mathrm{Det}(\mathbb{Z}_2\,\Delta^*_{n-1})}{K}\,. \tag{2.17}$$

With the notation of 3 (2.2), 3 (2.3)

$$\mathrm{Ker}(\mathbb{Z}\,\Gamma \to \Lambda) = \begin{cases} (1 + \sigma^{2^{n-2}})\,\mathbb{Z}\Gamma & \text{if } \Gamma = H_n, \\ (1 + \xi^{2^{n-2}})\,\mathbb{Z}\Gamma & \text{if } \Gamma = \Sigma_n\,. \end{cases}$$

Thus we see that

$$K \supseteq \mathrm{Hom}_{\Omega_Q}(R_{\Delta_{n-1}}, \mathcal{O}_F^*)\,\mathrm{Det}(1 + 2\mathbb{Z}_2\,\Delta_{n-1}). \tag{2.18}$$

Next we define

$$M_n = \frac{\mathrm{Hom}_{\Omega_Q}(R_{\Delta_{n-1}}, \mathcal{O}_F^*)\,\mathrm{Det}(\mathbb{Z}_2\,\Delta^*_{n-1})}{\mathrm{Hom}_{\Omega_Q}(R_{\Delta_{n-1}}, \mathcal{O}_F^*)\,\mathrm{Det}(1 + 2\mathbb{Z}\,\Delta_{n-1})}\,. \tag{2.19}$$

With the notation of 3 (2.1) we write

$$\Delta_{n-1} = \langle \alpha, \beta \;|\; \alpha^{2^{n-2}} = \beta^2 = 1, \;\; \beta\alpha\beta = \alpha^{-1} \rangle\,.$$

Thanks to Proposition 2.15, in order to prove Theorem (2.1.b), it will suffice to show:

Proposition 2.20

For $n \geq 3$, M_n is cyclic on $\mathrm{Det}(1 + \alpha + \beta)$.

Proof We argue by induction on n. When $n = 3$, Δ_2 is the Klein Vierer group and so M_3 is easily seen to be generated by $\mathrm{Det}(1 + \alpha + \beta)$. This starts our induction argument.

For $n > 3$, the homomorphism $\Delta_{n-1} \to \Delta_{n-2}$ induces a surjection $M_n \to M_{n-1}$. We let ϕ denote a faithful irreducible character of Δ_{n-1}. Let L denote the subgroup of those homomorphisms in the denominator of the right hand side of (2.19) which are identically 1 on the inflated image of $R_{\Delta_{n-2}}$. Evaluating on ϕ, we see that $\mathrm{Ker}(M_n \to M_{n-1})$ is a sub-quotient of

$$G = (\mathbb{Z}_2 [\eta_{n-1}]^+)^* / L(\phi).$$

However,

$$(1 - \alpha^{2^{n-3}}) \mathbb{Z}\Delta_{n-1} = \mathrm{Ker}(\mathbb{Z}\Delta_{n-1} \to \mathbb{Z}\Delta_{n-2})$$

and so L contains $\mathrm{Det}(1 + 2(1 - \alpha^{2^{n-3}}) \mathbb{Z}\Delta_{n-1})$, together with all those Ω_Q homomorphisms from $R_{\Delta_{n-1}}$ to $\mathcal{O}_F^*$ which are trivial on $\mathrm{Inf}_{\Delta_{n-2}}^{\Delta_{n-1}} R_{\Delta_{n-2}}$. On the other hand

$$\mathrm{Det}(1 + 2(1 - \alpha^{2^{n-3}})\mathbb{Z}_2 \langle\alpha\rangle)(\phi) = N(1 + 4\mathbb{Z}_2 [\eta_{n-1}]) \supseteq 1 + 4p,$$

by Lemma 2.3. We have therefore shown that G is a quotient of the group

$$G' = \frac{(\mathbb{Z}_2 [\eta_{n-1}]^+)^*}{(\mathbb{Z} [\eta_{n-1}]^+)^* \ (1 + 4p)} .$$

By (2.10) G′ has order at most 2 and is generated by the class of 5. However

$$\mathrm{Det}(1 + \alpha^{2^{n-4}} - \alpha^{-2^{n-4}})(\phi) = (1 + 2i)(1 - 2i) = 5.$$

Using the multiplicative congruence mod $1 + 2\mathbb{Z}_2\Delta_{n-1}$

$$1 + \alpha^{2^{n-4}} - \alpha^{-2^{n-4}} \equiv 1 + \alpha^{2^{n-4}} + \alpha^{-2^{n-4}}$$

$$\equiv (1 + \alpha + \alpha^{-1})^{2^{n-4}}$$

$$\equiv [(1 + \alpha + \beta)(1 + \alpha^{-1} + \beta)]^{2^{n-4}}$$

$$\equiv [(1 + \alpha + \beta)\beta(1 + \alpha + \beta)\beta]^{2^{n-4}}$$

we deduce that M_n is indeed cyclic on $\mathrm{Det}(1 + \alpha + \beta)$. □

<u>Proof of (2.1.c)</u> Let ψ be the irreducible non-abelian character of $_4H_3$, let χ_i, $1 \leq i \leq 4$, denote the abelian characters of H_3 and let $\rho = \sum_1 \chi_i$. We shall suppose χ_1 to be the identity character of H_3 and χ_2 to have kernel $\langle\sigma\rangle$ (with the notation of 3 (2.2)).

<u>Proposition 2.21</u>
For $z \in \mathbb{Z}_2 H_3^*$,

$$\mathrm{Det}(z)(\psi) \equiv (-1)^{\frac{1}{4}\log(\mathrm{Det}(z)(\rho))} \mod (4).$$

Before proving the proposition, we show that it implies (2.1.c).

From 1 (3.11)

$$D(\mathbb{Z} H_3) \cong \frac{\mathrm{Hom}_{\Omega_Q}(R_{H_3}, U_2(F))}{\mathrm{Hom}^+_{\Omega_Q}(R_{H_3}, \mathcal{O}_F^*)\ \mathrm{Det}(\mathbb{Z}_2 H_3^*)} . \tag{2.22}$$

(2.23) Moreover, because ψ is symplectic, the group $\mathrm{Hom}^+_{\Omega_Q}(R_{H_3}, \mathcal{O}_F^*)$ is the group of homomorphisms from R_{H_3} to ± 1 which are 1 on ψ.

We define a homomorphism

$$c_o: \mathrm{Hom}_{\Omega_Q}(R_{H_3}, U_2(F)) \to \pm 1 \mod (4)$$

by $c_o(g) = g(\psi).(-1)^{\frac{1}{4}\log(g(\rho))} \mod (4)$, for $g \in \mathrm{Hom}_{\Omega_Q}(R_{H_3}, U_2(F))$.

From (2.21) and (2.23) it follows that c_o maps the denominator in the right hand side of (2.22) to 1 mod (4). Hence c_o induces a homomorphism

$$c: D(\mathbb{Z} H_3) \to \pm 1 \bmod (4).$$

Recall that the class of the Swan module $[3, \Sigma_{H_3}]$ is represented by the homomorphism f_3, with the property that

$$f_3(\chi) = 3^{(\varepsilon,\chi)}$$

for $\chi \in R_{H_3}$. Thus $c\ (3,\Sigma_{H_3}) \equiv (-1)^{\frac{1}{4}\log(3)} \equiv -1 \bmod (4)$, which proves Theorem 2.1.c.

Proof of Proposition (2.21) From Proposition 1.4 (c), we have the congruence mod (4)

$$\begin{aligned} &\mathrm{Det}(z)(\psi) - \mathrm{Det}(z)(\chi_1) \\ \equiv\ &\frac{1}{2}\,[\mathrm{Det}(z)(\chi_1 + \chi_2) + \mathrm{Det}(z)(\chi_3 + \chi_4)] - \mathrm{Det}(z)(\chi_2). \end{aligned} \tag{2.24}$$

For $\theta \in R_{H_3}$ we put $\mathrm{Det}(z)(\theta) = 1 + 2x_\theta$. Note that since all characters of H_3 are rational valued, $x_\theta \in \mathbb{Z}_2$. We write x_i for x_{χ_i}. In order to prove the proposition it suffices to show

$$2x_\psi \equiv \sum_{i=1}^{4} x_i + x_i^2 \quad \bmod (4), \tag{2.25}$$

since $\log(\mathrm{Det}(z)(\rho)) = \log(\Pi(1 + 2x_i)) \equiv 2\sum_{i=1}^{4}(x_i - x_i^2 - 2x_i^4)$ mod (8), while $4x_i^4 \equiv 4x_i^2$ mod (8). Now (2.24) gives the congruence mod (4)

$$2x_\psi \equiv - x_1 - x_2 + x_3 + x_4 + 2x_1 x_2 + 2x_3 x_4. \tag{2.26}$$

However, this congruence depends on the particular choice of subgroup $\langle\sigma\rangle$ of index 2. Repeating for the remaining other two such groups, and summing, we obtain the congruence mod (4):

$$2x_\psi \equiv + x_1 + x_2 + x_3 + x_4 + \sum_{\substack{i,j=1\\ i>j}}^{4} 2x_i\, x_j \tag{2.27}$$

Let $x = \sum_1^4 x_i$. By (2.27) $x \equiv 0 \bmod (2)$, so that

$$- 2 \sum_{\substack{i,j\\ i>j}} x_i\, x_j \equiv 2 \sum_{\substack{i,j\\ i>j}} x_i\, x_j \equiv x^2 - \Sigma\, x_i^{\,2} \equiv - \Sigma\, x_i^{\,2} \bmod (4)$$

and so from (2.27)

$$2x_\psi \equiv \sum_{i=1}^{4} x_i + x_i^{\,2} \quad \bmod (4)$$

as was required. □

8. THE EXTENSION THEOREM FOR K_oT

§1. Let K be a tamely ramified Galois extension of Q_p and let $\Delta = \mathrm{Gal}(K/Q_p)$. Then $\otimes_{\mathbb{Z}_p} \mathcal{O}_K$ yields a homomorphism

$$\mathrm{Ind}_{Q_p}^{K} : K_oT(\mathbb{Z}_p\Gamma) \to K_oT(\mathcal{O}_K\Gamma).$$

In this chapter we show

<u>Theorem 1.1</u>

For K as above, the homomorphism $\mathrm{Ind}_{Q_p}^{K}$ is injective.

From 1 (4.13) we have the commutative square

$$\begin{array}{ccc}
K_oT(\mathcal{O}_K\Gamma) & \xrightarrow{\sim} & \dfrac{\mathrm{Hom}_{\Omega_K}(R_{\Gamma,p}, U_p)}{\mathrm{Det}(\mathcal{O}_K\Gamma^*)} \\
\uparrow {\scriptstyle \mathrm{Ind}_{Q_p}^{K}} & & \uparrow \\
K_oT(\mathbb{Z}_p\Gamma) & \xrightarrow{\sim} & \dfrac{\mathrm{Hom}_{\Omega_{Q_p}}(R_{\Gamma,p}, U_p)}{\mathrm{Det}(\mathbb{Z}_p\Gamma^*)}
\end{array}$$

We therefore obtain an isomorphism

$$\mathrm{Ker}(\mathrm{Ind}_{Q_p}^{K}) \cong \frac{\mathrm{Det}(\mathcal{O}_K\Gamma^*) \cap \mathrm{Hom}_{\Omega_{Q_p}}(R_{\Gamma,p}, U_p)}{\mathrm{Det}(\mathbb{Z}_p\Gamma^*)} .$$

Viewing $\mathrm{Det}(\mathcal{O}_K\Gamma^*)$ as an $\Omega_{\mathbb{Q}_p}$ (or Δ) module in the usual way, we have

$$\mathrm{Det}(\mathcal{O}_K\Gamma^*) \cap \mathrm{Hom}_{\Omega_{\mathbb{Q}_p}}(R_{\Gamma,p}, U_p) = \mathrm{Det}(\mathcal{O}_K\Gamma^*)^{\Delta}$$

where, as always, the superscript Δ denotes Δ-fixed points. We are therefore now reduced to showing:

Theorem 1.2

$$\mathrm{Det}(\mathcal{O}_K\Gamma^*)^{\Delta} = \mathrm{Det}(\mathbb{Z}_p\Gamma^*)$$

Remark Theorem 1.1, in this form, plays a fundamental role in describing rings of integers as Galois modules cf. [T3]. Indeed it was just this theorem which provided the motivation to reformulate the logarithmic techniques of [T2] into the group logarithm.

We will now show that in fact, it suffices to prove Theorem 1.2 when $K/\mathbb{Q}_p$ is non-ramified. Let L denote the maximal non-ramified extension of $\mathbb{Q}_p$ in K, and put $I = \mathrm{Gal}(K/L)$. Suppose $\mathrm{Det}(z) \in \mathrm{Det}(\mathcal{O}_K\Gamma^*)^{\Delta}$. Because K/L is totally ramified $\mathcal{O}_L\Gamma^*$ maps onto $\mathcal{O}_K\Gamma^*/_{1 + \pi\,\mathcal{O}_K\Gamma}$, where π generates the maximal ideal of $\mathcal{O}_K$. Thus we may write

$$\mathrm{Det}(z) = \mathrm{Det}(z_1).\mathrm{Det}(z_2) \qquad (1.3)$$

with $z_1 \in \mathcal{O}_L\Gamma^*$, $z_2 \in 1 + \pi\,\mathcal{O}_K\Gamma$. We now apply the norm map $N_{K/L}$ (cf. 1 (4.18)) and obtain

$$\mathrm{Det}(z\,z_1^{-1})^{|I|} = N_{K/L}\,(\mathrm{Det}(z\,z_1^{-1})) \in \mathrm{Det}(\mathcal{O}_L\Gamma^*).$$

However, from (1.3) we know $\mathrm{Det}(z\,z_1^{-1})$ lies in the pro-p-group $\mathrm{Det}(\mathcal{O}_L\Gamma^*) \cap \mathrm{Det}(1 + \pi\,\mathcal{O}_K\Gamma)$ and, of course, here raising to the power $|I|$, whence also $N_{K/L}$, is an automorphism since, by tameness, $(|I|,p) = 1$. Hence $\mathrm{Det}(z\,z_1^{-1}) \in \mathrm{Det}(\mathcal{O}_L\Gamma^*)$ and therefore

$$\mathrm{Det}(z) \in \mathrm{Det}(\mathcal{O}_L\Gamma^*).$$

For the remainder of this chapter we can now assume $K/\mathbb{Q}_p$ to be non-ramified. In fact, we shall prove the slightly more general result:

Theorem 1.4

Let $\Delta \subseteq \mathrm{Gal}(K/Q_p)$, then

$$\mathrm{Det}(\mathcal{O}_K\Gamma^*)^{\Delta} = \mathrm{Det}(\mathcal{O}_{K^{\Delta}}\Gamma^*).$$

Our proof proceeds in three stages. As a first step we show

Proposition 1.5

Let ℓ be a prime number and assume that Theorem 1.4 is true for all Q-ℓ-elementary groups Γ, then for any finite group Σ

$$\mathrm{Det}(\mathcal{O}_K\Sigma^*)^{\Delta}/\mathrm{Det}(\mathcal{O}_{K^{\Delta}}\Sigma^*)$$

has order coprime to ℓ.

Proof. From 12.6 in [Sel], we know that we can write

$$m\,\varepsilon_{\Sigma} = \Sigma\,\mathrm{Ind}_{\Gamma_i}^{\Sigma}(\theta_i) \tag{1.6}$$

where m is an integer coprime to ℓ, Γ_i ranges through the Q-ℓ-elementary subgroups of Σ and $\theta_i \in R_{\Gamma_i}(Q)$. (Note that the image of $R_{\Gamma_i}(Q)$ in $R_{\Gamma_i,p}$ is independent of choice of embedding $Q^c \hookrightarrow Q_p^c$).

Let $\mathrm{Det}(z) \in \mathrm{Det}(\mathcal{O}_K\Sigma^*)^{\Delta}$. From the character action formula 2 (2.1.c)

$$\mathrm{Det}(z)^m = \prod_i \mathrm{Ind}_{\Gamma_i}^{\Sigma}(\theta_i.\,\mathrm{Res}_{\Sigma}^{\Gamma_i}(\mathrm{Det}(z))).$$

Because $\mathrm{Res}_{\Sigma}^{\Gamma_i}$ and the action of $R_{\Gamma_i}(Q)$ take Δ-fixed homomorphisms to Δ-fixed homomorphisms, the result follows since our hypothesis implies that for each i

$$\mathrm{Det}(\mathcal{O}_K\Gamma_i^*)^{\Delta} = \mathrm{Det}(\mathcal{O}_{K^{\Delta}}\Gamma_i^*).$$

(1.7) The second step in the proof is to recall from 6 (3.12) that

Theorem 1.4 holds for all $\mathbb{Q}$-p-elementary groups.

For our third step we show that Theorem 1.4 holds for all $\mathbb{Q}$-ℓ-elementary groups with $\ell \neq p$. Theorem 1.4 will then hold in full generality by Proposition 1.5. We now suppose Γ to be $\mathbb{Q}$-ℓ-elementary with $\ell \neq p$. Hence Γ may be written

$$\Gamma = (C \times C') \rtimes P$$

where P is an ℓ-group and $C \times C'$ is a normal cyclic group with C (resp. C') of p-power order (resp. order prime to p). We let Γ' be the group $\Gamma/_C$.

Because $(|\Gamma'|,p) = 1$, the order $\mathcal{O}_K\Gamma'$ is maximal in $K\Gamma'$, and so from 1 (2.9)

$$\begin{aligned} \mathrm{Det}(\mathcal{O}_K\Gamma'^{*})^{\Delta} &= \mathrm{Hom}_{\Omega_K}(R_{\Gamma',p}, U_p)^{\Delta} \\ &= \mathrm{Hom}_{\Omega_{K^\Delta}}(R_{\Gamma',p}, U_p) \\ &= \mathrm{Det}(\mathcal{O}_{K^\Delta}\Gamma'^{*}) \end{aligned} \tag{1.8}$$

It is well known (cf. 8.2 in [Se1] for instance) that every irreducible character χ of Γ may be written in the form $\chi = \mathrm{Ind}_{\Xi}^{\Gamma}(\alpha)$, where α is an abelian character of Ξ and $\Xi \supset C \times C'$. We write α' for that part of α which has order coprime to p. Then $C \subset \mathrm{Ker}(\alpha')$, and so $\chi' = \mathrm{Ind}_{\Xi}^{\Gamma}\alpha'$ lies in $\mathrm{Inf}_{\Gamma'}^{\Gamma} R_{\Gamma'}$. Also, of course, $\chi - \chi' = \mathrm{Ind}_{\Xi}^{\Gamma}(\alpha - \alpha') \in \mathrm{Ker}\, d_{p,\Gamma}$.

Let $\mathrm{Det}(z) \in \mathrm{Det}(\mathcal{O}_K\Gamma^{*})^{\Delta}$, and let z map to z' under the algebra homomorphism $\mathcal{O}_K\Gamma \to \mathcal{O}_K\Gamma'$. By (1.8) there exists $x' \in \mathcal{O}_{K^\Delta}\Gamma'^{*}$ with the property that $\mathrm{Det}(z') = \mathrm{Det}(x')$. Moreover, because the kernel of $q: \mathcal{O}_{K^\Delta}\Gamma \to \mathcal{O}_{K^\Delta}\Gamma'$ is contained in the Jacobson radical of $\mathcal{O}_{K^\Delta}\Gamma$, we know q maps $\mathcal{O}_{K^\Delta}\Gamma^{*}$ onto $\mathcal{O}_{K^\Delta}\Gamma'^{*}$. Thus we can choose $x \in \mathcal{O}_{K^\Delta}\Gamma^{*}$ such that $q(x) = x'$.

(1.9) It is clear that $\mathrm{Det}(zx^{-1})$ is 1 on $\mathrm{Inf}_{\Gamma'}^{\Gamma} R_{\Gamma'}$. Furthermore by 4 (1.3)

$$\mathrm{Det}(zx^{-1})(\chi - \chi') \equiv 1 \bmod \mathcal{P}^{c}$$

so by (1.9)

$$\mathrm{Det}(zx^{-1})(\chi) \equiv 1 \bmod P^c.$$

It has therefore been shown that $\mathrm{Det}(zx^{-1}) \in \mathrm{Hom}_{\Omega_{K^\Delta}}(R_{\Gamma,p}, 1 + P^c)$, a pro-p-group. Since $\mathrm{Det}(\mathcal{O}_{K^\Delta}\Gamma^*)$ has finite index in $\mathrm{Hom}_{\Omega_{K^\Delta}}(R_{\Gamma,p}, U_p)$, we deduce

$$\mathrm{Det}(z)^{p^i} \in \mathrm{Det}(\mathcal{O}_{K^\Delta}\Gamma^*).$$

From Proposition 1.5 together with (1.7) it follows that

$$\mathrm{Det}(z) \in \mathrm{Det}(\mathcal{O}_{K^\Delta}\Gamma^*)$$

and therefore we have shown

$$\mathrm{Det}(\mathcal{O}_K\Gamma^*)^\Delta = \mathrm{Det}(\mathcal{O}_{K^\Delta}\Gamma^*)$$

for all Q-elementary groups. Thus again by Proposition 1.5 we have proved Theorem 1.4 in full generality.

9. ADAMS OPERATIONS FOR CLASSGROUPS

Let Γ be a finite group, let K be a finite, non-ramified extension of Q_p and let k be the residue classfield of $\mathcal{O}_K$. Let $G_oT(\mathcal{O}_K\Gamma)$ (resp. $G_o(k\Gamma)$) denote the Grothendieck group of finite $\mathcal{O}_K\Gamma$-modules taken modulo exact sequences (resp. of $k\Gamma$-modules taken modulo exact sequences). By taking composition factors, we obtain an isomorphism

$$G_oT(\mathcal{O}_K\Gamma) \cong G_o(k\Gamma). \tag{1}$$

In the sequel, for ease of notation, we shall identify $G_oT(\mathcal{O}_K\Gamma)$ and $G_o(k\Gamma)$. We assert that exterior products make $G_o(k\Gamma)$ into a λ-ring and in particular that the addition rule for λ-operations (see chapter 1, §2) is satisfied. Clearly exterior products induce maps on the isomorphism classes of $k\Gamma$ modules:

Lemma (cf. §2 of [Ke])

Let $0 \to V_o \to V_1 \overset{\pi}{\to} V_2 \to 0$ be an exact sequence of $k\Gamma$ modules. For $n \geq 0$, we have the equality in $G_o(k\Gamma)$

$$\lambda^n(V_1) = \sum_{i=o}^{n} \lambda^i(V_o)\, \lambda^{n-i}\,(V_2).$$

Proof It will suffice to exhibit a filtration of $k\Gamma$-modules

$$\lambda^n(V_1) = W_o \supset W_1 \quad . \quad . \supset W_n = \lambda^n(V_o) \supset W_{n+1} = (0)$$

with $W_i/W_{i+1} \cong \lambda^i(V_o) \otimes_k \lambda^{n-i}\,(V_2)$. We view V_o as a submodule of V_1, we let f denote the natural surjection $f\colon \otimes^n V_1 \to \lambda^n\,(V_1)$, and we set $W_i = f(V_o^i \otimes V_1^{n-i})$.

We consider the commutative diagram

$$\begin{array}{ccc} V_0{}^{i} \otimes V_1^{n-i} & \xrightarrow{\ f\ } & W_i \\ \Big\downarrow{\scriptstyle \pi'} & & \Big\downarrow{\scriptstyle q} \\ V_0{}^{i} \otimes V_2^{n-i} & \xrightarrow{\ f'\ } & W_i/W_{i+1} \end{array}$$

where f' is defined because $f(\mathrm{Ker}\ \pi') \subset f(V_0^{i+1} \otimes V_1^{n-i-1}) = W_{i+1}$. Moreover f' is onto since f and q are. We now obtain a commutative triangle:

$$\begin{array}{ccc} V_0{}^{i} \otimes V_2^{n-i} & \xrightarrow{\ f'\ } & W_i/W_{i+1} \\ & \searrow \quad \nearrow{\scriptstyle f^{\sim}} & \\ & \lambda^i(V_0) \otimes \lambda^{n-i}(V_2) & \end{array}$$

and the fact the $f^{\sim}$ are injective, and so isomorphisms, follows easily on counting dimensions. Namely, if $a = \dim_k V_0$, $b = \dim_k V_1$, $c = \dim_k V_2$, then $b = a + c$ and

$$\binom{b}{n} = \dim_k(\lambda^n(V_1)) = \sum_{i=1}^{n} \dim_k (W_i/W_{i+1})$$

$$\leq \sum \dim_k(\lambda^i(V_0)) \dim_k(\lambda^{n-i}(V_2)).$$

$$= \sum_i \binom{a}{i} \binom{c}{n-i} = \binom{b}{n}. \qquad \square$$

It now follows that the homomorphism $\lambda_t(V) = \sum_{i=0}^{\infty} \lambda^i(V)t^i$, for an indeterminate t, factorises through $G_0(k\Gamma)$. In particular the λ^i induce maps on $G_0(k\Gamma)$ and therefore, using Newton's formulae as in chapter 1, §2, we deduce the existence of Adams operations Ψ_h on $G_0(k\Gamma)$. By means of the isomorphism (1) we view the Ψ_h as endomorphisms of $G_0T(\mathcal{O}_K\Gamma)$.

In contrast, the exterior products of projective $\mathcal{O}_K\Gamma$ modules are not, in general, projective (just by consideration of rank). However, by use of the group logarithm we shall show that $K_0T(\mathcal{O}_K\Gamma)$ has natural Adams operations Ψ_h for each positive integer h, and, furthermore, that when $(h, |\Gamma|) = 1$, then the Ψ_h anti-commute with the Cartan map $c: K_0 T(\mathcal{O}_K\Gamma) \to G_0 T(\mathcal{O}_K\Gamma)$. That is to say, the diagram

$$\begin{array}{ccc} K_0 T(\mathcal{O}_K\Gamma) & \xrightarrow{c} & G_0 T(\mathcal{O}_K\Gamma) \\ \downarrow \Psi_j & & \downarrow \Psi_h \\ K_0 T(\mathcal{O}_K\Gamma) & \xrightarrow{c} & G_0 T(\mathcal{O}_K\Gamma) \end{array} \qquad (2)$$

commutes, when $jh \equiv 1 \bmod |\Gamma|$.

If N is a number field which is non-ramified at the prime divisors of $|\Gamma|$, then by the isomorphism 1 (3.1), $K_0 T(\mathcal{O}_N\Gamma)$ possesses Adams operations, and we are therefore able to use the exact sequence 1 (1.3) to deduce the structure of an Adams ring on $K_0(\mathcal{O}_N\Gamma)$ (cf. (1.12)).

When Γ is abelian our results are elementary and have been known for some time. In this case these Adams operations have been used to view classgroups as modules over the Iwasawa algebra and so obtain a fine description of such classgroups cf. [KM], [U5].

The results of this chapter are joint work of the author and Philippe Cassou-Noguès, and they can be seen as the natural analogue of the work of M. Kervaire [Ke].

§1. STATEMENT OF RESULTS

Composition with the endomorphism $\psi_h: R_{\Gamma,p} \to R_{\Gamma,p}$ yields an enomorphism

$$\Psi_h: \mathrm{Hom}_{\Omega_K}(R_{\Gamma,p}, \mathbb{Q}_p^{c*}) \to \mathrm{Hom}_{\Omega_K}(R_{\Gamma,p}, \mathbb{Q}_p^{c*}). \qquad (1.1)$$

In sections 3 to 5 we prove:

__Theorem 1.2__

$$\Psi_h(\mathrm{Det}(\mathcal{O}_K\Gamma^*)) \subseteq \mathrm{Det}(\mathcal{O}_K\Gamma^*).$$

The Ψ_h therefore induce endomorphisms, which we also denote Ψ_h, on $K_0T(\mathcal{O}_K\Gamma)$, via the isomorphism 1 (3.2).

Theorem 1.3

For all positive integers h

(1) $$\Psi_h \circ \Psi_{h'} = \Psi_{hh'}$$

(2) If e is the exponent of Γ, then on $K_0T(\mathcal{O}_K\Gamma)$

$$\Psi_{h+e} = \Psi_h .$$

(3) The diagram (2) commutes.

Properties (1) and (2) follow from the corresponding properties for the Adams operations on $R_{\Gamma,p}$. We now set out to establish property (3). So we now fix a positive integer h, coprime to $|\Gamma|$ and a positive integer j such that $hj \equiv 1 \bmod |\Gamma|$. We show that for $x \in K_0T(\mathcal{O}_K\Gamma)$

$$\Psi_j \circ c(x) = c \circ \Psi_h(x). \tag{1.4}$$

Without loss of generality we assume k to be 'big enough'. [Here we are using the fact that if k′ is an extension of k, then $G_0(k\Gamma) \to G_0(k'\Gamma)$ is a homomorphism of λ-rings and is injective (cf. (14.6) in [Sel])].

In order to prove (1.4) we need a Fröhlich-type description of $G_0(k\Gamma)$ due to Queyrut which is based on the pairing between $G_0(k\Gamma)$ and $P_0(k\Gamma)$ (the Grothendieck group of projective $k\Gamma$-modules)

$$\langle\ ,\ \rangle_k : P_0(k\Gamma) \times G_0(k\Gamma) \to \mathbb{Z}$$

induced by the map $(P,M) \mapsto \dim_k(\mathrm{Hom}_{k\Gamma}(P,M))$.

We shall write

$$\langle\ ,\ \rangle_K : G_0(K\Gamma) \times G_0(K\Gamma) \to \mathbb{Z}$$

for the standard inner product of $K\Gamma$ characters.

Theorem 1.5 (cf. §3 [Q])

Let $H(k,\Gamma)$ be the subgroup of those homomorphisms in $\mathrm{Hom}_{\Omega_K}(R_{\Gamma,p}, Q_p^{c*})$ which on Im(e) take values in U_p, where $e: P_0^K(k\Gamma) \to G_0(K\Gamma)$ is the usual homomorphism in the cde Swan triangle (cf. §15 [Sel]). Then

(1) $$G_0(k\Gamma) \cong \frac{\mathrm{Hom}_{\Omega_K}(R_{\Gamma,p}, Q_p^{c*})}{H(k,\Gamma)}$$

(2) The following square commutes:

$$\begin{array}{ccc} K_0T(\mathcal{O}_K\Gamma) & \tilde{\to} & \dfrac{\mathrm{Hom}_{\Omega_K}(R_{\Gamma,p}, Q_p^{c*})}{\mathrm{Det}(\mathcal{O}_K\Gamma^*)} \\ \Big\downarrow c & & \Big\downarrow \\ G_0(k\Gamma) & \tilde{\to} & \dfrac{\mathrm{Hom}_{\Omega_K}(R_{\Gamma,p}, Q_p^{c*})}{H(k,\Gamma)} \end{array}$$

(3) A $k\Gamma$-module M is represented by a homomorphism f under the isomorphism (1), if, and only if,

$$\langle P,M\rangle = v(f(e(P))),$$

for all projective $k\Gamma$-modules P where v is that valuation of K which is 1 on the maximal ideal of $\mathcal{O}_K$.

$P(k\Gamma)$ identifies as an ideal of $G_0(k\Gamma)$ via the Cartan map c. Kervaire has shown that $P(k\Gamma)$ is stable under the operations Ψ_h (cf. [Ke]). We may therefore view the Ψ_h as operations on $P(k\Gamma)$ (with the property that $c \circ \Psi_h = \Psi_h \circ c$).

Lemma 1.6

For $x \in P(k\Gamma)$, $\psi_h \circ e(x) = e \circ \Psi_h(x)$

Proof We let $y = \psi_h \circ e(x) - e \circ \Psi_h(x)$. Since $c = d \circ e$

$$d(y) = d \circ \psi_h \circ e(x) - c \circ \Psi_h(x).$$

However, we know that $c \circ \Psi_h = \psi_h \circ c$ and, of course $d \circ \psi_h = \Psi_h \circ d$ since the decomposition map d is a homomorphism of λ-rings. Thus $y \in \mathrm{Ker}(d)$. From Theorem 36 in [Sel], we know that $\mathrm{Im}(e)$ is ψ_h stable in $G_o(K\Gamma)$. Thus $y \in \mathrm{Im}(e)$, and therefore $y \in \mathrm{Ker}(d) \cap \mathrm{Im}(e) = (o)$.

We now prove (1.4). From part (3) of Theorem 1.5 we must show that for each projective class $y \in P(k\Gamma)$ and for $x \in K_oT(\mathcal{O}_K\Gamma)$

$$\langle y, \Psi_h \circ c(x)\rangle_k = \langle y, c \circ \Psi_j(x)\rangle_k. \qquad (1.7)$$

Suppose $x \in K_oT(\mathcal{O}_K\Gamma)$ is represented by $f \in \mathrm{Hom}_{\Omega_K}(R_{\Gamma,p}, Q_p^{c^*})$. By definition $\Psi_h(x)$ is represented by the homomorphism $f \circ \psi_h$ and so by part (2) of Theorem 1.5, $c(\Psi_h(x))$ is also represented by $f \circ \psi_h$. In the same way $c(x)$ is represented by f. By part (3) of Theorem 1.5, we are now reduced to showing that if $w \in G_o(k\Gamma)$ is represented by g, then

$$\langle y, \Psi_h(w)\rangle_k = v(g \circ \psi_j \circ e(y)) \qquad (1.8)$$

for all $y \in P(k\Gamma)$. Now choose $z \in G_o(K\Gamma)$ such that $d(z) = w$. Let $e(y)$ have character θ and let z have character χ. Because e and d are adjoint, and since ψ_h commutes with both e (by Lemma 1.6) and d (being a homomorphism of λ-rings):

$$\langle y, \Psi_h \circ d(z)\rangle = \langle y, d \circ \psi_h(z)\rangle$$

$$= \langle e(y), \psi_h(z)\rangle_K$$

$$= \frac{1}{|\Gamma|} \sum_{\gamma\in\Gamma} \chi(\gamma^h)\ \theta(\gamma^{-1})$$

since $(h, |\Gamma|) = 1$

$$= \frac{1}{|\Gamma|} \sum_{\gamma\in\Gamma} \chi(\gamma)\ \theta(\gamma^{-j})$$

$$= \langle \psi_j e(y), z\rangle_K$$

$$= \langle e \circ \Psi_j(y), z\rangle_K$$

$$= \langle \Psi_j(y), d(z)\rangle_k$$

$$= \langle \Psi_j(y), w \rangle_k$$

$$= v(g \circ e \circ \Psi_j(y)) = v(g \circ \psi_j \circ e(y))$$

as required. □

We now turn to the global situation and we let N be a number field which is non-ramified at the prime divisors of $|\Gamma|$. We say that a positive integer h is <u>suitable</u> for Γ if either h is odd or if Γ has no irreducible symplectic characters.

For any positive integer h, by means of the isomorphism 1 (3.1), the above Adams operations Ψ_h, induce an endomorphism, again denoted Ψ_h, on $K_oT(\mathcal{O}_N\Gamma)$. In order to obtain a corresponding endomorphism of $Cl(\mathcal{O}_N\Gamma)$, we must consider when $\theta(K_1(N\Gamma))$ is Ψ_h stable (cf. 1 (1.3)).

We now re-express this in terms of groups of homomorphisms. By 1 (2.10) we now wish to establish when $Hom^+_{\Omega_N}(R_\Gamma, F^*)$ is Ψ_h stable, where, by abuse of notation, Ψ_h also denotes the endomorphism of $Hom_{\Omega_N}(R_\Gamma, J(F))$ given by composition with $\psi_h: R_\Gamma \to R_\Gamma$.

<u>Proposition 1.9</u>

If h is suitable for Γ, then $Hom^+_{\Omega_N}(R_\Gamma,F^*)$ is Ψ_h stable.

<u>Proof</u> If Γ has no irreducible symplectic characters then $Hom^+_{\Omega_N}(R_\Gamma,F^*) = Hom_{\Omega_N}(R_\Gamma,F^*)$ and the result is immediate. So now let h be odd and let R^s_Γ denote the subgroup of R_Γ generated by the symplectic characters of Γ. In order to prove the proposition it suffices to show:

<u>Lemma 1.10</u>

If h is odd, then $\psi_h(R^s_\Gamma) \subseteq R^s_\Gamma$.

<u>Remark</u> If h is even, then, in general, R^s_Γ is not ψ_h stable.

<u>Proof</u> We begin by showing that it suffices to take Γ to be real elementary, i.e. $\Gamma = \Sigma \rtimes P$ with Σ cyclic and normal, P a p-group $(|\Sigma|,|P|) = 1$ and

$$\pi^{-1} \sigma \pi = \sigma^{\pm 1} \qquad (1.11)$$

for $\pi \in P$, $\sigma \in \Sigma$.

From Proposition 36 in [Sel], we may write

$$\varepsilon_\Gamma = \Sigma \operatorname{Ind}_{\Gamma_i}^{\Gamma} \theta_i$$

$\theta_i \in R^o_{\Gamma_i}$, the orthogonal virtual characters of Γ_i, where the sum extends over all real elementary subgroups Γ_i of Γ. By Frobenius reciprocity, for $\chi \in R^s_\Gamma$,

$$\psi_h\chi = \Sigma \operatorname{Ind}_{\Gamma_i}^{\Gamma} (\theta_i\, \psi_h(\operatorname{Res}_\Gamma^{\Gamma_i} \theta_i))$$

Since $\operatorname{Ind}_{\Gamma_i}^{\Gamma}(R^s_{\Gamma_i}) \subseteq R^s_\Gamma$ and $R^s_{\Gamma_i}\, R^o_{\Gamma_i} \subseteq R^s_\Gamma$, we see that it suffices to take Γ real elementary.

If $|P|$ is odd, then, by (1.11), C and P commute, so that the only symplectic characters are of degenerate type i.e., of the form $\phi + \bar{\phi}$. Hence the result in this case is immediate.

We now suppose P to be a 2-group. It suffices to show $\psi_h\chi$ is symplectic when χ is an irreducible symplectic character of Γ. However, being irreducible, χ must be of the form

$$\chi = \operatorname{Ind}_{\Sigma H}^{\Gamma} \xi.\eta$$

where ξ is an abelian character of Σ, H is the stabiliser of ξ and η is an irreducible character of H. By Lemma 3.1 in section 3

$$\psi_h\chi = \operatorname{Ind}_{\Sigma H}^{\Gamma} \xi^h.\psi_h\eta.$$

The map $\sigma\pi \mapsto \sigma^h\pi$ for $\sigma \in \Sigma$, $\pi \in P$ is an endomorphism of Γ, so that $\operatorname{Ind}_{\Sigma H}^{\Gamma} \xi^h\eta$ must be symplectic, since χ is. It remains only to observe that $\operatorname{Ind}_{\Sigma H}^{\Gamma} \xi^h\eta$ is a Galois conjugate of $\psi_h\chi$. □

(1.12) We now extend the endomorphism Ψ_h of $Cl(\mathcal{O}_N\Gamma)$, for suitable h, to $K_o(\mathcal{O}_N\Gamma)$, by defining $\Psi_h([\mathcal{O}_N\Gamma]) = [\mathcal{O}_N\Gamma]$ and by linearity. We then obtain endomorphisms Ψ_h of $K_o(\mathcal{O}_N\Gamma)$ for suitable h, which satisfy conditions (1) and (2) of Theorem 1.3. In fact the Ψ_h are ring endomorphisms of $K_o(\mathcal{O}_N\Gamma)$. Because the free classes are an ideal in $K_o(\mathcal{O}_N\Gamma)$, we need only check that for c, d $\in Cl(\mathcal{O}_K\Gamma)$

$$\Psi_h(c)\ \Psi_h(d) = \Psi_h(cd).$$

However, both sides are seen to be zero, since, by 2 (1.3), for $c \in Cl(\mathcal{O}_N\Gamma)$ and x a locally free $\mathcal{O}_N\Gamma$ module, c.x is zero.

We conclude this section by remarking that the analogue to Theorem 1.3 (3) is an open question: namely, for h coprime to $|\Gamma|$, do the hth Adams operations commute with the Cartan map from $K_o(\mathcal{O}_N\Gamma)$ to the Grothendieck group of $\mathcal{O}_N$-projective, $\mathcal{O}_N\Gamma$-modules?

§2. REDUCTION TO Q-p-ELEMENTARY GROUPS

We now turn to the heart of the matter which is Theorem 1.2. Our notation is as in §1. Clearly it suffices to prove when h is a prime number. Define

$$M_{K,\Gamma} = \frac{\Psi_h(\mathrm{Det}(\mathcal{O}_K\Gamma^*))\mathrm{Det}(\mathcal{O}_K\Gamma^*)}{\mathrm{Det}(\mathcal{O}_K\Gamma^*)} \qquad (2.1)$$

The proof of Theorem 1.2 is, in many ways, similar in structure to that of 8 Theorem 1.4.

We identify $R_\Gamma(Q)$, the ring of Q-characters, with its image in $R_{\Gamma,p}$.

Proposition 2.2

Let ℓ be a prime number and assume $M_{K,\Gamma} = 1$ for all Q-ℓ-elementary groups Γ. Then $M_{K,\Sigma}$ has order prime to ℓ, for all finite groups Σ.

Proof Again we may write

$$m\ \varepsilon_\Sigma = \sum_i \mathrm{Ind}_{\Gamma_i}^{\Sigma}\ \theta_i$$

where $(m,\ell) = 1$ and Γ_i ranges through the Q-ℓ-elementary subgroups of Σ. For $\chi \in R_\Sigma$

$$m.\psi_h\ \chi = \sum_i \mathrm{Ind}_{\Gamma_i}^{\Sigma}\ (\theta_i.\ \mathrm{Res}_{\Sigma}^{\Gamma_i}\ (\psi_h\ \chi)),$$

and so for $z \in \mathcal{O}_K\Sigma^*$

$$(\Psi_h \operatorname{Det}(z))^m = \prod_i \operatorname{Det}(z)(\operatorname{Ind}_{\Gamma_i}^{\Sigma}(\theta_i \cdot \psi_h(\operatorname{Res}_{\Sigma}^{\Gamma_i} \chi)))$$

$$= \prod_i \Psi_h \operatorname{Det}(z_i)(\operatorname{Res}_{\Sigma}^{\Gamma_i} \chi)$$

where $\operatorname{Det}(z_i) = \theta_i \operatorname{Res}_{\Sigma}^{\Gamma_i}(\operatorname{Det}(z))$. Thus

$$(\Psi_h \operatorname{Det}(z))^m = \prod_i \operatorname{Ind}_{\Gamma_i}^{\Sigma}(\Psi_h (\operatorname{Det}(z_i)))$$

and, of course, $\Psi_h(\operatorname{Det}(z_i)) \in \operatorname{Det}(\mathcal{O}_K \Gamma_i^*)$, by hypothesis.

The second (and main) step in the proof of Theorem 1.2, is:

Proposition 2.3

If Γ is Q-p-elementary, then $M_{K,\Gamma} = 1$.

This is where all the difficulty lies, and, in order that the structure of the proof be as clear as possible, we postpone the proof of Proposition (2.3) until the next section.

From Proposition (2.2) and (2.3) we deduce

Corollary 2.4

$M_{K,\Gamma}$ has order prime to p, for all finite groups Γ.

The third and final step in the proof of (1.2) is:

Proposition 2.5

If Γ is Q-ℓ-elementary with $\ell \neq p$, then $M_{K,\Gamma}$ has p-power order.

Clearly Proposition (2.5) and corollary (2.6) imply that $M_{K,\Gamma} = 1$ for all Q-elementary groups; therefore by Proposition 2.3 we have shown Theorem 1.2.

Proof of Proposition 2.5 We remark that the proof is very close to that of step 3 in the proof of 8 Theorem 1.4. So, for brevity, Γ, Γ', χ, χ', $\equiv$, q are all taken to be as used in that proof. We start by proving the proposition for Γ'. Since $(|\Gamma'|,p) = 1$, $\mathcal{O}_K\Gamma'$ is a maximal order in $K\Gamma'$ and so

$$\operatorname{Det}(\mathcal{O}_K\Gamma') = \operatorname{Hom}_{\Omega_K}(R_{\Gamma',p}, U_p).$$

It is immediate that the right hand side is Ψ_h stable.

Now we prove the result for Γ. Let $z \in \mathcal{O}_K\Gamma^*$. By the above there exists $x_o \in \mathcal{O}_K\Gamma'^*$ such that

$$\Psi_h(\mathrm{Det}(q(z))) = \mathrm{Det}(x_o).$$

Since $q(\mathcal{O}_K\Gamma^*) = \mathcal{O}_K\Gamma'^*$, we may choose $x \in \mathcal{O}_K\Gamma^*$ such that $q(x) = x_o$. We consider

$$f = \Psi_h(\mathrm{Det}(z))\,\mathrm{Det}(x^{-1}).$$

Clearly f is 1 on $\mathrm{Inf}_{\Gamma'}^{\Gamma}(R_{\Gamma',p})$. Furthermore, because $\chi - \chi'$, whence also $\psi_h(\chi - \chi')$, lies in Ker $d_{p,\Gamma}$; from 4 (1.3)

$$f(\chi - \chi') \equiv 1 \bmod P^c.$$

whence

$$f(\chi) \equiv 1 \bmod P^c$$

since $\chi' \in \mathrm{Inf}_{\Gamma'}^{\Gamma}(R_{\Gamma',p})$. Therefore we have shown that

$$M_{K,\Gamma} \hookrightarrow \frac{\mathrm{Hom}_{\Omega_K}(R_{\Gamma,p}, 1 + P^c)\,\mathrm{Det}(\mathcal{O}_K\Gamma^*)}{\mathrm{Det}(\mathcal{O}_K\Gamma^*)}.$$

Since the right hand group is finite and $1 + P^c$ is a pro-p-group, it follows that $M_{K,\Gamma}$ must have p-power order.

§3. PROOF OF (2.3)

The idea behind this proof is to reduce from Q-p-elementary groups to p-groups and then apply the techniques of chapter 6. We start with the following elementary result:

Lemma 3.1

Let Δ be a normal sub-group of Γ and let $(m,(\Gamma:\Delta)) = 1$. Then for $\alpha \in R_\Delta$

$$\psi_m(\mathrm{Ind}_\Delta^\Gamma\ \alpha) = \mathrm{Ind}_\Delta^\Gamma\ \psi_m\alpha.$$

In particular, when m is a prime number the operator δ_m commutes with $\mathrm{Ind}_\Delta^\Gamma$.

<u>Proof</u> Because $\gamma \in \Delta$ if, and only if, $\gamma^m \in \Delta$, the result is an immediate consequence of the formula for an induced character evaluated on a group element. □

We now recall some notation and results from chapter 6 §3. Let $\Gamma = \Sigma \rtimes P$ be a Q-p-elementary group, with Σ normal, cyclic and of order coprime to p, and with P a p-group.

Let m be a divisor of $|\Sigma|$ and the $\xi\colon \Sigma \to Q_p^{c^*}$ be an abelian character with order m. We write H_m for the kernel of the homomorphism $P \to \mathrm{Aut}(\xi(\Sigma))$ induced by conjugating P on Σ. We set $A_m = P/H_m$, $\Gamma_m = \Sigma\, H_m$, and let K(m) denote the field obtained by adjoining the mth roots of unity in Q_p^c to K. For the time being we view m as fixed and so we surpress the subscript m on A and H. Let $R_H^{(m)}$ denote the group of virtual characters generated by the characters of the form $\mathrm{Ind}_{\Gamma_m}^\Gamma\ (\chi\theta)$ where χ is an abelian character of Σ with order m, $\theta \in R_H$. We recall that $R_P^{(m)} = \mathrm{Ind}_{\Gamma_m}^\Gamma\ (R_H^{(m)})$.

From 6 (3.6), (3.8), (3.9) we have a commutative square

$$\begin{array}{ccc} \mathrm{Hom}_{\Omega_K}(R_{\Gamma,p}, U_p) & \to & \prod_m \mathrm{Hom}_{\Omega_K}(R_H^{(m)}, U_p)^{A_m} \\ \uparrow & & \uparrow \\ \mathrm{Det}(\mathcal{O}_K\Gamma^*) & \tilde{\to} & \prod_m \mathrm{Det}(\mathcal{O}_K[m]H^*)^{A_m} \end{array} \qquad (3.2)$$

Now let $\xi_{m,1} \cdots \xi_{m,k}$ be representatives for the Ω_K orbits of abelian characters of order m. Then

$$R_H^{(m)} = \sum_1^k \xi_{m,i}\, R_H \cdot \mathbb{Z}\Omega_K .$$

We let $\nu_{m,i}$ be the surjection

$$\nu_{m,i} : \mathrm{Hom}_{\Omega_K}(R_H^{(m)}, U_p) \to \mathrm{Hom}_{\Omega_{K(m)}}(R_{H,p}, U_p)$$

given by

$$(\nu_{m,i} f)(\theta) = f(\xi_{m,i} \cdot \theta)$$

for $\theta \in R_{H_m}$. Then $\prod_1^k \nu_{m,i}$ yields an isomorphism, ν say,

$$\nu\colon \mathrm{Hom}_{\Omega_K}(R_H^{(m)}, U_p) \cong \prod_1^k \mathrm{Hom}_{\Omega_{K(m)}}(R_{H,p}, U_p). \tag{3.3}$$

It is easily seen that

$$\nu(\mathrm{Det}(\mathcal{O}_K[m]H^*) = \prod_1^k \mathrm{Det}(\mathcal{O}_{K(m)}H^*). \tag{3.4}$$

From (3.2) we now obtain a further commutative square

$$\begin{array}{ccc} \mathrm{Hom}_{\Omega_K}(R_{\Gamma,p}, U_p) & \overset{\alpha}{\hookrightarrow} & \prod_m \left(\prod_1^{k(m)} \mathrm{Hom}_{\Omega_{K(m)}}(R_{H,p}, U_p)\right)^{A_m} \\ \uparrow & & \uparrow \\ \mathrm{Det}(\mathcal{O}_K\Gamma^*) & \xrightarrow[\sim]{\alpha'} & \prod_m \left(\prod_1^{k(m)} \mathrm{Det}(\mathcal{O}_{K(m)} H^*)\right)^{A_m} . \end{array} \tag{3.5}$$

<u>Remark</u> It should be pointed out that the A_m action on the right of (3.5) is that deduced from the isomorphism of (3.4). The A_m action on the left of (3.3) and (3.4) is given in 6 (3.7).

Writing $\alpha_{m,i}$ for the composite of α and projection onto the (m,i) th factor, we see that for $f \in \mathrm{Hom}_{\Omega_K}(R_{\Gamma,p}, U_p)$

$$f \in \mathrm{Det}(\mathcal{O}_K\Gamma^*) \iff \alpha_{m,i}(f) \in \mathrm{Det}(\mathcal{O}_{K(m)} H^*) \tag{3.6}$$

for all permissible pairs m,i. Explicitly, for $\theta \in R_{H,p}$

$$(\alpha_{m,i}(f))(\theta) = f(\mathrm{Ind}_{\Gamma_m}^{\Gamma}(\xi_{m,i}\theta)). \tag{3.7}$$

The reduction from Q-p-elementary groups to p-groups is considerably less straightforward than usual, because, in general, Adams operations fail to commute with induction. We are able to get around this difficulty by means of the following result:

Lemma 3.8

With the above notation, where ξ is an abelian character of Σ of order m and $\theta \in R_{H,p}$

$$\left(\prod_{q|m}(1-\delta_q)\right)\left[\psi_p(\mathrm{Ind}_{\Gamma_m}^{\Gamma}\xi.\theta) - \mathrm{Ind}_{\Gamma_m}^{\Gamma}(\psi_p(\xi.\theta))\right] = 0.$$

Here the product extends over all prime divisors q of m.

Proof For $x \in \Sigma$, $y \in P$ with the property that either $(xy)^p \notin \Gamma_m$ or $xy \in \Gamma_m$, it is immediate that for any $\alpha \in R_{\Gamma_m}$

$$\psi_p(\mathrm{Ind}_{\Gamma_m}^{\Gamma}\alpha)(xy) = \mathrm{Ind}_{\Gamma_m}^{\Gamma}(\psi_p\alpha)(xy).$$

In these two cases the result follows by Lemma (3.1). So now let us suppose

$$x \in \Sigma,\ y \in P\backslash H_m \quad \text{and} \quad y^p \in H_m \ . \tag{3.9}$$

We let σ be a generator of Σ. Then for some integer r prime to $|\Sigma|$

$$y^{-1}\sigma y = \sigma^r .$$

From (3.9), since Γ_m is the stabiliser of ξ, we may immediately deduce

$$r \not\equiv 1 \bmod (m) \qquad r^p \equiv 1 \bmod (m). \tag{3.10}$$

Furthermore, because $(m,p) = 1$, we see that, if $r \equiv 1 \bmod q_o$ for a prime q_o with $q_o^{n_o} \| m$, then $r \equiv 1 \bmod q_o^{n_o}$. This remark, together with (3.9), implies the existence of a prime q_o dividing m such that

$$r \not\equiv 1 \bmod q_o \qquad r^p \equiv 1 \bmod q_o^{n_o} \tag{3.11}$$

where $q_o^{n_o}$ is the exact power of q_o dividing m.

Writing $s = (r^p - 1)(r - 1)^{-1}$, we have

$$(xy)^p = x^s\ y^p.$$

Hence, using the formula for an induced character, we see that for x, y as in (3.9)

$$\psi_p(\mathrm{Ind}_{\Gamma_m}^{\Gamma}\ \xi.\theta)(xy) = \sum_{\alpha\in H_m\setminus P} \xi(\alpha^{-1}x\alpha)^s\ \theta(\alpha^{-1}y^p\alpha).$$

From (3.11) $\xi(\alpha^{-1}x\alpha)^s = \delta_{q_0}\ \xi(\alpha^{-1}x^s\alpha)$, and of course, $\delta_{q_0}\ \theta(\alpha^{-1}y^p\alpha) = \theta(\alpha^{-1}y^p\alpha)$ since y has p-power order. We have therefore shown, in this case, that

$$\psi_p(\mathrm{Ind}_{\Gamma_m}^{\Gamma}\ \xi\theta)(xy) = \mathrm{Ind}_{\Gamma_m}^{\Gamma}\ (\delta_{q_0}\ (\xi\theta))(x^s y^p)).$$

Since δ_{q_0} is idempotent and commutes with $\mathrm{Ind}_{\Gamma_m}^{\Gamma}$, we see that

$$(\prod_{q|m}\ (1 - \delta_q))\ \psi_p(\mathrm{Ind}_{\Gamma_m}^{\Gamma}\ \ \xi.\theta)(xy) = 0.$$

Again since the δ_q commute with $\mathrm{Ind}_{\Gamma_m}^{\Gamma}$

$$(\prod_{q|m}\ (1 - \delta_q))\ \mathrm{Ind}_{\Gamma_m}^{\Gamma}\ (\xi.\theta) = \mathrm{Ind}_{\Gamma_m}^{\Gamma}\ \ \alpha$$

for some $\alpha \in R_{\Gamma_m}$ and this clearly vanishes on such xy. □

In addition to Lemma 3.8 we also need the following elementary result:

<u>Lemma 3.12</u>

Let h be an integer and let $a \in \mathcal{O}_K\Gamma^*$. Then there exists $b \in \mathcal{O}_K\Gamma^*$ such that for all $\alpha \in R_\Sigma$, $\theta \in R_{H_m}$ for all $m \,\big|\, |\Sigma|$

$$\mathrm{Det}(b)(\mathrm{Ind}_{\Gamma_m}^{\Gamma}\ \ \alpha.\theta) = \mathrm{Det}(a)(\mathrm{Ind}_{\Gamma_m}^{\Gamma}\ (\psi_h\alpha).\theta)$$

<u>Proof</u> For $\sigma \in \Sigma$, $\pi \in P$, the map $\sigma\pi \mapsto \sigma^h\pi$ is a group endomorphism $t_h\colon \Gamma \to \Gamma$. Extending t_h to a group algebra homomorphism and setting $b = t_h(a)$ gives the result. □

For the present assume the following result (which we prove at the end of this section).

Proposition 3.13

For all p-groups P and for all non-ramified extensions L/Q_p, $M_{L,P} = 1$.

We start the proof of Proposition (2.3) by considering the case when the prime number h is different from p. From (3.6) it suffices to show that for $a \in \mathcal{O}_K\Gamma^*$

$$\alpha_{m,i}\,(\Psi_h\,(\mathrm{Det}(a))) \in \mathrm{Det}(\mathcal{O}_{K(m)}\,H^*) \qquad (3.14)$$

for all pairs (m,i) as given in (3.5). For brevity we denote $\xi_{m,i}$ (resp. $\alpha_{m,i}$) by ξ (resp. α). From (3.7), we see that for $\theta \in R_{H,p}$

$$\alpha(\Psi_h(\mathrm{Det}(a)))(\theta) = \mathrm{Det}(a)(\psi_h\ \mathrm{Ind}_{\Gamma_m}^{\Gamma}\ \xi.\theta)$$

by Lemma 3.1

$$= \mathrm{Det}(a)(\mathrm{Ind}_{\Gamma_m}^{\Gamma}\ \xi^h.\psi_h\theta).$$

Thus, with the notation of Lemma 3.12,

$$\alpha(\Psi_h(\mathrm{Det}(a)))(\theta) = \mathrm{Det}(b)(\mathrm{Ind}_{\Gamma_m}^{\Gamma}\ \xi.\psi_h\theta)$$

$$= \Psi_h(\alpha(\mathrm{Det}(b))(\theta).$$

However, by (3.6) $\alpha(\mathrm{Det}(b)) \in \mathrm{Det}(\mathcal{O}_{K(m)}H^*)$ and by Proposition 3.13 this is Ψ_h stable. Therefore $\alpha(\Psi_h(\mathrm{Det}(a))) \in \mathrm{Det}(\mathcal{O}_{K(m)}H^*)$. □

Now we consider the case when h = p. Here we argue by induction on $w(\Sigma)$, the number of prime divisors of Σ. If $w(\Sigma) = 0$, then Γ is a p-group and the result is true by Proposition 3.13. So now suppose $w(\Sigma) > 0$, and let $\mathcal{P}$ denote the set of non-empty sets of prime divisors of $|\Sigma|$. For $A \in \mathcal{P}$ and $\chi \in R_\Sigma$, let

$$\chi_A = (\prod_{\ell \in A} \delta_\ell)\chi$$

From Lemma 3.8, we know that

$$\psi_p(\mathrm{Ind}_{\Gamma_m}^{\Gamma}\ \xi.\theta) = \mathrm{Ind}_{\Gamma_m}^{\Gamma}\ (\psi_p(\xi\theta)) + \sum_{A \in \mathcal{P}} \mu_A(\psi_p(\mathrm{Ind}_{\Gamma_m}^{\Gamma}\ \xi_A\ \theta) \qquad (3.15)$$

$$- \mathrm{Ind}_{\Gamma_m}^{\Gamma}\ \psi_p(\xi_A\ \theta))$$

where $\mu_A = (-1)^{1+\text{card}(A)}$. Let $\Sigma(A)$ be the quotient group of Σ by the subgroup of elements whose orders contain only primes in A. Let $\Gamma(A) = \Sigma(A) \rtimes P$. We view $\Gamma(A)$ as a quotient of Γ. For each $A \in \mathcal{P}$ the virtual characters $\psi_p(\mathrm{Ind}_{\Gamma_m}^{\Gamma}(\xi_A.\theta))$, $\mathrm{Ind}_{\Gamma_m}^{\Gamma}\psi_p(\xi_A\theta)$ both lie in $\mathrm{Inf}_{\Gamma(A)}^{\Gamma}(R_{\Gamma(A),p})$. Therefore, by the induction hypothesis, because $w(\Sigma(A)) < w(\Sigma)$, there exists $a_1 \in \mathcal{O}_K\Gamma(A)^*$ such that

$$\mathrm{Det}(a)(\psi_p\ \mathrm{Ind}_{\Gamma(A)_m}^{\Gamma(A)}(\xi_A\ \theta)) = \mathrm{Det}(a_1)(\mathrm{Ind}_{\Gamma_m}^{\Gamma}(\xi_A\ \theta)).$$

Here $\Gamma(A)_m$ is the image of Γ_m in $\Gamma(A)$. Since $\mathcal{O}_K\Gamma^*$ maps onto $\mathcal{O}_K\Gamma(A)^*$, we may choose $c_A \in \mathcal{O}_K\Gamma^*$ which maps onto a_1. Thus

$$\mathrm{Det}(a)(\psi_p\ \mathrm{Ind}_{\Gamma_m}^{\Gamma}(\xi_A\ \theta)) = \mathrm{Det}(c_A)(\mathrm{Ind}_{\Gamma_m}^{\Gamma}(\xi_A.\theta))\ . \tag{3.16}$$

Next we use Lemma 3.12 to deduce the existence of elements d, b_A in $\mathcal{O}_K\Gamma^*$ such that

$$\mathrm{Det}(a)(\mathrm{Ind}_{\Gamma_m}^{\Gamma}\ \psi_p(\xi.\theta)) = \mathrm{Det}(d)(\mathrm{Ind}_{\Gamma_m}^{\Gamma}\ \xi.\psi_p\theta)$$

$$\mathrm{Det}(a)(\mathrm{Ind}_{\Gamma_m}^{\Gamma}\ \psi_p(\xi_A\ \theta)) = \mathrm{Det}(b_A)(\mathrm{Ind}_{\Gamma_m}^{\Gamma}\ \xi.\psi_p\theta)$$

for all $\theta \in R_{H,p}$. From (3.7) we re-interpret this as

$$\begin{aligned}\mathrm{Det}(a)(\mathrm{Ind}_{\Gamma_m}^{\Gamma}\ \psi_p(\xi.\theta)) &= \Psi_p(\alpha(\mathrm{Det}(d))(\theta),\\ \mathrm{Det}(a)(\mathrm{Ind}_{\Gamma_m}^{\Gamma}\ \psi_p(\xi_A.\theta)) &= \Psi_p(\alpha(\mathrm{Det}(b_A))(\theta).\end{aligned} \tag{3.17}$$

So, from (3.15), (3.16) and (3.17),

$$\alpha(\Psi_p(\mathrm{Det}(a)) = \Psi_p(\alpha(\mathrm{Det}(d)) \prod_A \alpha(\mathrm{Det}(c_A^{\mu_A}))\Psi_p \circ \alpha(\mathrm{Det}(b_A^{-\mu_A})).$$

However, by Proposition 3.13, we know that the right hand side is an element of $\mathrm{Det}(\mathcal{O}_{K(m)}\ H^*)$, and so, by (3.6), $\Psi_p(\mathrm{Det}(a)) \in \mathrm{Det}(\mathcal{O}_K\Gamma^*)$.

Finally we must prove Proposition (3.13). So now let Γ be a p-group and henceforth, without further remark, we use the notation of Chapter 6, §1.

Let $a = \sum_{\gamma\in\Gamma} a_\gamma\ \gamma$ with $a_\gamma \in \mathcal{O}_K$ and suppose $a \in \mathcal{O}_K\Gamma^*$. Let

$b = \Sigma a_\gamma \gamma^h$. Since the Jacobson radical is stable under $\gamma \mapsto \gamma^h$, $b \in \mathcal{O}_K\Gamma^*$. We set $g = \Psi_h \operatorname{Det}(a).\operatorname{Det}(b^{-1})$. It is clear that for an abelian character χ of Γ

$$\operatorname{Det}(a)(\psi_h\chi) = \operatorname{Det}(b)(\chi),\ g(\chi) = 1. \tag{3.18}$$

Next consider the image of g under the homomorphism ν_o (cf. 6 (1.6)).

Lemma 3.19

$$\nu_o(g) \in p\mathcal{O}_K C.$$

Proof Let $\nu(\operatorname{Det}(a)) = \sum_{c\in C} x_c\ c(\gamma)$; then $x_c \in p\mathcal{O}_K$ by 6 (1.2.b). Likewise $\nu(\operatorname{Det}(b)) \in p\mathcal{O}_K C$; therefore it will suffice to show

$$\nu_o(\Psi_h(\operatorname{Det}(a))) = \Sigma x_c\ c(\gamma^h).$$

Since the irreducible characters of Γ span $\operatorname{Hom}_{Q_p^c}(Q_p^c C, Q_p^c)$, it suffices to prove that for $\chi \in R_{\Gamma,p}$

$$\chi(\nu_o(\Psi_h(\operatorname{Det}(a)))) = \chi(\Sigma x_c\ c(\gamma^h)).$$

Now by the definition of ψ_h

$$\chi(\Sigma x_c\ c(\gamma^h)) = \psi_h\chi(\Sigma x_c\ c(\gamma))$$

$$= \psi_h\chi(\nu(\operatorname{Det}(a)))$$

and so, by 6 (1.6),

$$\chi(\Sigma x_c\ c(\gamma^h)) = \log(\operatorname{Det}(a)(\psi_h\chi)) - \log(\operatorname{Det}(a^f)(\psi_p \circ \psi_h\chi))$$

$$= \chi(\nu_o(\Psi_h \operatorname{Det}(a)))$$

since $\psi_p \circ \psi_h = \psi_{ph} = \psi_h \circ \psi_p$. This proves the lemma. □

By (3.18) and 6 (1.6), $\chi(\nu_o(g)) = 0$ for all abelian characters of Γ. Hence $\nu_o(g) \in K.\phi(a_K)$ (recall $a_K = \mathrm{Ker}(\mathcal{O}_K\Gamma \to \mathcal{O}_K\Gamma^{ab})$), and so

$$\nu_o(g) \in p\mathcal{O}_K C \cap K.\phi(a_K).$$

However, $\phi(a_K)$ is easily seen to be a direct summand of $\mathcal{O}_K C$ (as $\mathcal{O}_K$-modules), so that

$$p\mathcal{O}_K C \cap K\phi(a_K) = p\phi(a_K).$$

By 6 (1.12) and 6 (1.7)

$$g = \mathrm{Det}(1 + z)t \tag{3.20}$$

for some $z \in a_K$ and where t lies in the subgroup T_K' of Ω_K homomorphisms from R_Γ to p-power roots of unity in Q_p^c, which are 1 on all abelian characters of Γ. To see that t has p-power order (and not just finite order), let $\chi \in R_{\Gamma,p}$ and let $d = \chi(1)$ be its degree. Now t is of the form

$$t = \Psi_h(\mathrm{Det}(a)).\mathrm{Det}(x)$$

for some $x \in \mathcal{O}_K\Gamma^*$. Therefore

$$t(\chi) = t(\chi - d\varepsilon_\Gamma) = \mathrm{Det}(a)(\psi_h(\chi - d\varepsilon_\Gamma))\mathrm{Det}(x)(\chi - d\varepsilon_\Gamma). \tag{3.21}$$

Since $\chi - d\varepsilon_\Gamma$, whence also $\psi_h(\chi - d\varepsilon_\Gamma)$, lies in Ker $d_{p,\Gamma}$, the right hand term in (3.21) lies in $1 + P^c$ cf. 4 (1.3).

We wish to use the decomposition (3.20) to define a monomorphism

$$\pi : M_{K,\Gamma} \hookrightarrow T_K' \tag{3.22}$$

by $\pi(\Psi_h(\mathrm{Det}(a))\ \mathrm{Det}(\mathcal{O}_K\Gamma^*)) = t$. The injectivity of π will be immediate once we show π to be well-defined. However, from 5 (1.1) we know that

$$\mathrm{Det}(\mathcal{O}_K\Gamma^*) \cap T_K' = \{1\}$$

and so π is indeed well-defined.

Let T'_K have exponent p^e. We then denote by L the unique non-ramified extension of K of degree p^e. We remark that since any p^e th roots of unity of L lies in Q_p : $T'_L = T'_K = T'_{Q_p}$. Moreover, for $t \in T'_L$

$$\mathcal{N}_{L/K} t = t^{[L:K]} = t^{p^e} = 1. \qquad (3.23)$$

<u>Proposition 3.24</u>

$$\mathcal{N}_{L/K} \operatorname{Det}(\mathcal{O}_L \Gamma^*) = \operatorname{Det}(\mathcal{O}_K \Gamma^*).$$

Hence, because $\mathcal{N}_{L/K}$ and Ψ_h commute, $\mathcal{N}_{L/K}$ induces a surjection from $M_{L,\Gamma}$ onto $M_{K,\Gamma}$.

Before proving the proposition, we observe that we have a commutative diagram with exact rows:

$$\begin{array}{ccccc} 0 & \longrightarrow & M_{L,\Gamma} & \xrightarrow{\pi} & T'_{Q_p} \\ & & \downarrow{\scriptstyle \mathcal{N}_1} & & \downarrow{\scriptstyle \mathcal{N}_2} \\ 0 & \longrightarrow & M_{K,\Gamma} & \xrightarrow{\pi} & T'_{Q_p} \\ & & \downarrow & & \\ & & 0 & & \end{array}$$

Since $\mathcal{N}_2$ is null, $\pi \circ \mathcal{N}_1$ and whence $\mathcal{N}_1$ is null: therefore $M_{K,\Gamma} = 1$. This concludes our proof of Proposition 3.13.

<u>Proof of Proposition 3.24</u> The second part will follow immediately from the first. The exact sequence 6 (1.21) and $\mathcal{N}_{L/K}$ gives rise to a commutative diagram with exact rows:

$$\begin{array}{ccccccc}
0 \to p\phi(\mathcal{A}_{\mathbb{Q}_p}) \otimes_{\mathbb{Z}_p} \mathcal{O}_L & \longrightarrow & \mathrm{Det}(\mathcal{O}_L\Gamma^*) & \longrightarrow & \mathcal{O}_L\Gamma^{ab*} & \longrightarrow & 1 \\
\downarrow \mathrm{Tr}_{L/K} & & \downarrow N_{L/K} & & \downarrow N_{L/K} & & \\
0 \to p\phi(\mathcal{A}_{\mathbb{Q}_p}) \otimes_{\mathbb{Z}_p} \mathcal{O}_K & \longrightarrow & \mathrm{Det}(\mathcal{O}_K\Gamma^*) & \longrightarrow & \mathcal{O}_K\Gamma^{ab*} & \longrightarrow & 1
\end{array}$$

Here:

(1) $N_{L/K}$ is induced by the usual algebra norm $L\Gamma^{ab} \to K\Gamma^{ab}$. Because L/K is non-ramified, $N_{L/K}$ is surjective (cf. 6 (3.10)).

(2) $\mathrm{Tr}_{L/K}$ is induced by the trace tr: $L \to K$.
Again since L/K is non-ramified $\mathrm{tr}(\mathcal{O}_L) = \mathcal{O}_K$ and so $\mathrm{Tr}_{L/K}$ is surjective. □

REFERENCES

[B] H. Bass, "Algebraic K-theory", Mathematics Lecture Notes Series, Benjamin, New York, 1968.

[CN1] Ph. Cassou-Noguès, Structure galoisienne des anneaux d'entiers, Proc. LMS (3) 38, 1979, 545-576.

[CN2] Ph. Cassou-Noguès, Classes d'idéaux de l'algèbre d'un groupe abélien, C.R. Acad. Sci. Paris, 276 (1973), 973-975.

[F1] A. Fröhlich, Arithmetic and Galois module structure for tame extensions, J. für Math., 286/7 (1976), 380-440.

[F2] A. Fröhlich, Classgroups, in particular Hermitian classgroups, to appear.

[F3] A. Fröhlich, A normal integral basis theorem, J. of Algebra 39, 1976, 131-137.

[F4] A. Fröhlich, Local fields, in Algebraic Number Theory (ed. Cassels and Fröhlich), Academic Press.

[F5] A. Fröhlich, On the classgroup of integral group rings of finite abelian groups I, Mathematika, 16 (1969), 143-152.

[F6] A. Fröhlich, On the classgroup of integral group rings of finite abelian groups II, Mathematika, 19 (1972), 51-56.

[F7] A. Fröhlich, Galois module structure of algebraic integers, Springer Ergebnisse, 3 Folge, Band 1, 1983.

[FKW] A. Fröhlich, M.E. Keating, S.M.J. Wilson, The classgroup of quaternion and dihedral 2-groups, Mathematika, 21 (1974), 64-71.

[G] S. Galovich, The classgroup of a cyclic p-group, J. of Algebra, 30, 1974, 368-387.

[GRU] S. Galovich, I. Reiner, S. Ullom, Classgroups for integral representations of metacyclic groups, Mathematika, 19, (1972), 105-111.

[He] E. Hecke, Vorlesungen über die Theorie der algebraischen Zahlen, Chelsea (1948).

[J] H. Jacobinski, Two remarks about hereditary orders, Proc. Am. Math. Soc, 28 (1971), 1-8.

[Kt] M.E. Keating, Classgroups of metacyclic groups of order $p^r q$, p a regular prime, Mathematika 21, (1974), 90-95.

[Ke] M.A. Kervaire, Opérations d'Adams en théorie des représentations linéaires des groupes finis, Ens. Math., 22 (1976), 1-28.

[KM] M.A. Kervaire, M.P. Murthy, On the projective classgroup of cyclic groups of prime power order, Comment. Math. Helv. 52 (1977), 415-452.

[L] T.Y. Lam, Artin exponent of finite groups, J. of Algebra, 9, (1968), 94-119.

[Mc] L. Maculloh, Galois module structure of elementary abelian extensions, to appear.

[O1] R. Oliver, SK_1 for finite group rings II, Math. Scand 47, 1980, 195-231.

[O2] R. Oliver, $D(\mathbb{Z}\pi)^+$ and the Artin cokernel, Aarhus preprint.

[Q] J. Queyrut, S-groupes des classes d'un ordre arithmétique, to appear in J. of Algebra.

[RU] I. Reiner, S. Ullom, A Mayer-Vietoris sequence for classgroups, J. of Algebra, 31 (1974) 305-342.

[Re] I. Reiner, Maximal orders, Academic Press, London, 1975.

[R] D.S. Rim, Modules over finite groups, Annals of Math. 69 (1959), 700-712.

[Se1] J.P. Serre, Représentations linéaires des groupes finis, 2nd edition, Hermann, Paris 1971.

[Se2] J.P. Serre, Corps locaux, Hermann, Paris, 1968.

[Sw1] R.G. Swan, Periodic resolutions for finite groups, Ann. of Math, 72, (1960) 267-291.

[Sw2] R.G. Swan, Induced representations of projective modules, Ann. of Math. 71 (1960), 552-578.

[SE] R.G. Swan, E.G. Evans, K-theory of finite groups and orders, Springer Lecture Notes 149, 1970.

[T1] M.J. Taylor, A logarithmic approach to classgroups of integral group rings, J. of Algebra, 66, (1980), 321-353.

[T2] M.J. Taylor, Locally free classgroups of prime power order, J. of Algebra, 50 (1978) 462-487.

[T3] M.J. Taylor, On Fröhlich's conjecture for rings of integers of tame extensions, Invent. Math. 63 (1981), 41-79.

[T4] M.J. Taylor, On the self-duality of a ring of integers as a Galois module, Invent. Math. 46 (1978), 173-177.

[U1] S. Ullom, Character action on the classgroup of Fröhlich, in Algebraic K-theory, ed. R.K. Denmis, Springer Lecture Notes 967, 1981.

[U2] S. Ullom, The exponent of classgroups, J. of Algebra, 29 (1974), 124-132.

[U3] S. Ullom, A survey of integral group rings, Algebraic Number Fields, ed. A. Fröhlich, Academic Press.

[U4] S. Ullom, Non-trivial lower bounds for classgroups of integral group rings, Illinois J. Math. 20 (1976), 361-371.

[U5] S. Ullom, Fine structure of classgroups of cyclic p-groups, J. of Algebra, 49 (1977), 112-124.

[W1] C.T.C. Wall, Norms of units in group rings, Proc. London Math. Soc. (3), 29, (1974), 593-632.

[W2] C.T.C. Wall, Periodic projective resolutions, Proc. London Math. Soc. (3) (1979) 509-553.

[We] H. Weber, Lehrbuch der Algebra, 2nd edition, (2), Chelsea.

[Wi] S.M.J. Wilson, K-theory for twisted group rings, Proc. London Math. Soc. (3) 29 (1974) 257-271.

[Z] H. Zassenhaus, Theory of groups, Chelsea (1958).